OpenCV 机器学习(影印版)
Machine Learning for OpenCV

Michael Beyeler 著

南京　东南大学出版社

图书在版编目(CIP)数据

OpenCV 机器学习:英文/(美)迈克尔·贝耶勒(Michael Beyeler)著. —影印本. —南京:东南大学出版社,2019.5

书名原文:Machine Learning for OpenCV
ISBN 978-7-5641-8324-0

Ⅰ.①O… Ⅱ.①迈… Ⅲ.①机器学习-算法-英文 Ⅳ.①TP181

中国版本图书馆 CIP 数据核字(2019)第 046190 号
图字:10-2018-502 号

© 2017 by PACKT Publishing Ltd.

Reprint of the English Edition, jointly published by PACKT Publishing Ltd and Southeast University Press, 2019. Authorized reprint of the original English edition, 2018 PACKT Publishing Ltd, the owner of all rights to publish and sell the same.

All rights reserved including the rights of reproduction in whole or in part in any form.

英文原版由 PACKT Publishing Ltd 出版 2017。

英文影印版由东南大学出版社出版 2019。此影印版的出版和销售得到出版权和销售权的所有者—— PACKT Publishing Ltd 的许可。

版权所有,未得书面许可,本书的任何部分和全部不得以任何形式重制。

OpenCV 机器学习(影印版)

出版发行:东南大学出版社
地　　址:南京四牌楼 2 号　　邮编:210096
出 版 人:江建中
网　　址:http://www.seupress.com
电子邮件:press@seupress.com
印　　刷:常州市武进第三印刷有限公司
开　　本:787 毫米×980 毫米　　16 开本
印　　张:23.75
字　　数:465 千字
版　　次:2019 年 5 月第 1 版
印　　次:2019 年 5 月第 1 次印刷
书　　号:ISBN 978-7-5641-8324-0
定　　价:96.00 元

本社图书若有印装质量问题,请直接与营销部联系。电话(传真):025-83791830

Credits

Author
Michael Beyeler

Reviewers
Vipul Sharma
Rahul Kavi

Commissioning Editor
Veena Pagare

Acquisition Editor
Varsha Shetty

Content Development Editor
Jagruti Babaria

Technical Editor
Sagar Sawant

Copy Editor
Manisha Sinha

Project Coordinator
Manthan Patel

Proofreader
Safis Editing

Indexer
Tejal Daruwale Soni

Graphics
Tania Dutta

Production Coordinator
Deepika Naik

Foreword

Over the last few years, our machines have slowly but surely learned how to see for themselves. We now take it for granted that our cameras detect our faces in pictures that we take, and that social media apps can even recognize us and our friends in the photos that we upload from these cameras. Over the next few years we will experience even more radical transformation. Before long, cars will be driving themselves, our cellphones will be able to read and translate a sign in any language for us, and our x-rays and other medical images will be read and analyzed by powerful algorithms that will be able to accurately suggest a medical diagnosis, and even recommend effective treatments.

These transformations are driven by an explosive combination of increased computing power, masses of image data, and a set of clever ideas taken from math, statistics, and computer science. This rapidly growing intersection that is machine learning has taken off, affecting many of our day-to-day interactions with the world, and with each other. One of the most remarkable features of the current machine learning paradigm-shift in computer vision is that it relies to a large extent on software tools that are freely available and developed by large groups of volunteers, hobbyists, scientists, and engineers in open source communities. This means that, in principle, the barriers to entry are also lower than ever: anyone who is interested in putting their mind to it can harness machine learning for image processing.

However, just like in a garden with many forking paths, the wealth of tools and ideas, and the rapid development of these ideas, underscores the need for a guide who can show you the way, and orient you in the right direction. I have some good news for you: having picked up this book, you are in the good hands of my colleague and collaborator Dr. Michael Beyeler as your guide. With his broad range of expertise, Michael is both a hard-nosed engineer, computer scientist, and neuroscientist, as well as a prolific open source software developer. He has not only taught robots how to see and navigate through complex environments, and computers how to model brain activity, but he also regularly teaches humans how to use programming to solve a variety of different machine learning and image processing problems. This means that you will get to benefit not only from the sure-handed rigor of his expertise and experience, but also that you will get to enjoy his thoughtfulness in teaching the ideas in his book, as well as a good dose of his sense of humor.

The second piece of good news is that this going to be an exhilarating trip. There's nothing that matches the thrill of understanding that comes from putting together the pieces of the puzzle that go into solving a problem in computer vision and machine learning with code and data. As Richard Feynman put it: "What I cannot create, I do not understand". So, get ready to get your hands dirty (so to speak) with the code and data in the (open source!) code examples that accompany this book, and to get creative. Understanding will surely follow.

Ariel Rokem
Data Scientist, The University of Washington eScience Institute

About the Author

Michael Beyeler is a Postdoctoral Fellow in Neuroengineering and Data Science at the University of Washington, where he is working on computational models of bionic vision in order to improve the perceptual experience of blind patients implanted with a retinal prosthesis (bionic eye). His work lies at the intersection of neuroscience, computer engineering, computer vision, and machine learning. Michael is the author of *OpenCV with Python Blueprints* by Packt Publishing, 2015, a practical guide for building advanced computer vision projects. He is also an active contributor to several open source software projects, and has professional programming experience in Python, C/C++, CUDA, MATLAB, and Android.

Michael received a PhD in computer science from the University of California, Irvine as well as a MSc in biomedical engineering and a BSc in electrical engineering from ETH Zurich, Switzerland. When he is not "nerding out" on brains, he can be found on top of a snowy mountain, in front of a live band, or behind the piano.

About the Reviewers

Vipul Sharma is a Software Engineer at a startup in Bangalore, India. He studied engineering in Information Technology at Jabalpur Engineering College (2016). He is an ardent Python fan and loves building projects on computer vision in his spare time. He is an open source enthusiast and hunts for interesting projects to contribute to. He is passionate about learning and strives to better himself as a developer. He writes blogs on his side projects at `http://vipul.xyz`. He also publishes his code at `http://github.com/vipul-sharma20`.

Rahul Kavi works as a research scientist in Silicon Valley. He holds a Master's and PhD degree in computer science from West Virginia University. Rahul has worked on researching and optimizing computer vision applications for a wide variety of platforms and applications. He has also contributed to the machine learning module in OpenCV. He has written computer vision and machine learning software for prize-winning robots for NASA's 2015 and 2016 Centennial Challenges: Sample Return Robot (1st prize). Rahul's research has been published in conference papers and journals.

www.PacktPub.com

For support files and downloads related to your book, please visit www.PacktPub.com.

Did you know that Packt offers eBook versions of every book published, with PDF and ePub files available? You can upgrade to the eBook version at www.PacktPub.com and as a print book customer, you are entitled to a discount on the eBook copy. Get in touch with us at service@packtpub.com for more details.

At www.PacktPub.com, you can also read a collection of free technical articles, sign up for a range of free newsletters and receive exclusive discounts and offers on Packt books and eBooks.

https://www.packtpub.com/mapt

Get the most in-demand software skills with Mapt. Mapt gives you full access to all Packt books and video courses, as well as industry-leading tools to help you plan your personal development and advance your career.

Why subscribe?

- Fully searchable across every book published by Packt
- Copy and paste, print, and bookmark content
- On demand and accessible via a web browser

Customer Feedback

Thanks for purchasing this Packt book. At Packt, quality is at the heart of our editorial process. To help us improve, please leave us an honest review on this book's Amazon page at https://www.amazon.com/dp/1783980281.

If you'd like to join our team of regular reviewers, you can e-mail us at customerreviews@packtpub.com. We award our regular reviewers with free eBooks and videos in exchange for their valuable feedback. Help us be relentless in improving our products!

To my loving wife, who continues to support me in all my endeavors – no matter how grand, silly, or nerdy they may be.

Table of Contents

Preface	1
Chapter 1: A Taste of Machine Learning	9
Getting started with machine learning	10
Problems that machine learning can solve	12
Getting started with Python	14
Getting started with OpenCV	14
Installation	15
Getting the latest code for this book	15
Getting to grips with Python's Anaconda distribution	16
Installing OpenCV in a conda environment	19
Verifying the installation	19
Getting a glimpse of OpenCV's ML module	22
Summary	23
Chapter 2: Working with Data in OpenCV and Python	25
Understanding the machine learning workflow	26
Dealing with data using OpenCV and Python	29
Starting a new IPython or Jupyter session	29
Dealing with data using Python's NumPy package	31
Importing NumPy	31
Understanding NumPy arrays	32
Accessing single array elements by indexing	34
Creating multidimensional arrays	35
Loading external datasets in Python	36
Visualizing the data using Matplotlib	38
Importing Matplotlib	38
Producing a simple plot	38
Visualizing data from an external dataset	40
Dealing with data using OpenCV's TrainData container in C++	43
Summary	45
Chapter 3: First Steps in Supervised Learning	47
Understanding supervised learning	48
Having a look at supervised learning in OpenCV	49
Measuring model performance with scoring functions	50
Scoring classifiers using accuracy, precision, and recall	50
Scoring regressors using mean squared error, explained variance, and R squared	54

Using classification models to predict class labels — 58
Understanding the k-NN algorithm — 59
Implementing k-NN in OpenCV — 60
Generating the training data — 60
Training the classifier — 64
Predicting the label of a new data point — 65
Using regression models to predict continuous outcomes — 68
Understanding linear regression — 68
Using linear regression to predict Boston housing prices — 70
Loading the dataset — 71
Training the model — 71
Testing the model — 72
Applying Lasso and ridge regression — 74
Classifying iris species using logistic regression — 76
Understanding logistic regression — 76
Loading the training data — 77
Making it a binary classification problem — 78
Inspecting the data — 79
Splitting the data into training and test sets — 80
Training the classifier — 80
Testing the classifier — 81
Summary — 82

Chapter 4: Representing Data and Engineering Features — 83
Understanding feature engineering — 84
Preprocessing data — 85
Standardizing features — 86
Normalizing features — 87
Scaling features to a range — 88
Binarizing features — 88
Handling the missing data — 89
Understanding dimensionality reduction — 91
Implementing Principal Component Analysis (PCA) in OpenCV — 93
Implementing Independent Component Analysis (ICA) — 97
Implementing Non-negative Matrix Factorization (NMF) — 98
Representing categorical variables — 100
Representing text features — 102
Representing images — 104
Using color spaces — 104
Encoding images in RGB space — 104
Encoding images in HSV and HLS space — 106
Detecting corners in images — 106

        Using the Scale-Invariant Feature Transform (SIFT) 108
        Using Speeded Up Robust Features (SURF) 110
    Summary 111

Chapter 5: Using Decision Trees to Make a Medical Diagnosis — 113

    Understanding decision trees 114
        Building our first decision tree 117
            Understanding the task by understanding the data 119
            Preprocessing the data 123
            Constructing the tree 124
        Visualizing a trained decision tree 125
        Investigating the inner workings of a decision tree 127
        Rating the importance of features 128
        Understanding the decision rules 130
        Controlling the complexity of decision trees 131
    Using decision trees to diagnose breast cancer 132
        Loading the dataset 133
        Building the decision tree 134
    Using decision trees for regression 139
    Summary 142

Chapter 6: Detecting Pedestrians with Support Vector Machines — 143

    Understanding linear support vector machines 144
        Learning optimal decision boundaries 144
        Implementing our first support vector machine 146
            Generating the dataset 147
            Visualizing the dataset 147
            Preprocessing the dataset 148
            Building the support vector machine 149
            Visualizing the decision boundary 150
    Dealing with nonlinear decision boundaries 153
        Understanding the kernel trick 154
        Knowing our kernels 155
        Implementing nonlinear support vector machines 156
    Detecting pedestrians in the wild 158
        Obtaining the dataset 158
        Taking a glimpse at the histogram of oriented gradients (HOG) 160
        Generating negatives 162
        Implementing the support vector machine 165
        Bootstrapping the model 166
        Detecting pedestrians in a larger image 167
        Further improving the model 170

Summary	171
Chapter 7: Implementing a Spam Filter with Bayesian Learning	**173**
Understanding Bayesian inference	174
Taking a short detour on probability theory	174
Understanding Bayes' theorem	176
Understanding the naive Bayes classifier	180
Implementing your first Bayesian classifier	181
Creating a toy dataset	181
Classifying the data with a normal Bayes classifier	183
Classifying the data with a naive Bayes classifier	186
Visualizing conditional probabilities	187
Classifying emails using the naive Bayes classifier	189
Loading the dataset	190
Building a data matrix using Pandas	193
Preprocessing the data	194
Training a normal Bayes classifier	195
Training on the full dataset	195
Using n-grams to improve the result	196
Using tf-idf to improve the result	197
Summary	198
Chapter 8: Discovering Hidden Structures with Unsupervised Learning	**199**
Understanding unsupervised learning	200
Understanding k-means clustering	201
Implementing our first k-means example	201
Understanding expectation-maximization	204
Implementing our own expectation-maximization solution	205
Knowing the limitations of expectation-maximization	207
First caveat: No guarantee of finding the global optimum	208
Second caveat: We must select the number of clusters beforehand	209
Third caveat: Cluster boundaries are linear	212
Fourth caveat: k-means is slow for a large number of samples	215
Compressing color spaces using k-means	215
Visualizing the true-color palette	215
Reducing the color palette using k-means	219
Classifying handwritten digits using k-means	222
Loading the dataset	222
Running k-means	222
Organizing clusters as a hierarchical tree	225
Understanding hierarchical clustering	225

Implementing agglomerative hierarchical clustering	226
Summary	227

Chapter 9: Using Deep Learning to Classify Handwritten Digits — 229

Understanding the McCulloch-Pitts neuron	230
Understanding the perceptron	233
Implementing your first perceptron	236
Generating a toy dataset	237
Fitting the perceptron to data	238
Evaluating the perceptron classifier	239
Applying the perceptron to data that is not linearly separable	241
Understanding multilayer perceptrons	243
Understanding gradient descent	244
Training multi-layer perceptrons with backpropagation	248
Implementing a multilayer perceptron in OpenCV	249
Preprocessing the data	250
Creating an MLP classifier in OpenCV	250
Customizing the MLP classifier	251
Training and testing the MLP classifier	253
Getting acquainted with deep learning	256
Getting acquainted with Keras	257
Classifying handwritten digits	260
Loading the MNIST dataset	260
Preprocessing the MNIST dataset	262
Training an MLP using OpenCV	263
Training a deep neural net using Keras	264
Preprocessing the MNIST dataset	265
Creating a convolutional neural network	266
Fitting the model	267
Summary	268

Chapter 10: Combining Different Algorithms into an Ensemble — 269

Understanding ensemble methods	270
Understanding averaging ensembles	272
Implementing a bagging classifier	272
Implementing a bagging regressor	274
Understanding boosting ensembles	275
Implementing a boosting classifier	276
Implementing a boosting regressor	277
Understanding stacking ensembles	278
Combining decision trees into a random forest	279
Understanding the shortcomings of decision trees	279

Implementing our first random forest	284
Implementing a random forest with scikit-learn	286
Implementing extremely randomized trees	288
Using random forests for face recognition	290
Loading the dataset	290
Preprocessing the dataset	291
Training and testing the random forest	293
Implementing AdaBoost	295
Implementing AdaBoost in OpenCV	295
Implementing AdaBoost in scikit-learn	297
Combining different models into a voting classifier	298
Understanding different voting schemes	298
Implementing a voting classifier	299
Summary	301

Chapter 11: Selecting the Right Model with Hyperparameter Tuning — 303

Evaluating a model	304
Evaluating a model the wrong way	304
Evaluating a model in the right way	305
Selecting the best model	307
Understanding cross-validation	310
Manually implementing cross-validation in OpenCV	313
Using scikit-learn for k-fold cross-validation	314
Implementing leave-one-out cross-validation	315
Estimating robustness using bootstrapping	316
Manually implementing bootstrapping in OpenCV	317
Assessing the significance of our results	319
Implementing Student's t-test	320
Implementing McNemar's test	322
Tuning hyperparameters with grid search	324
Implementing a simple grid search	324
Understanding the value of a validation set	325
Combining grid search with cross-validation	328
Combining grid search with nested cross-validation	330
Scoring models using different evaluation metrics	332
Choosing the right classification metric	332
Choosing the right regression metric	333
Chaining algorithms together to form a pipeline	334
Implementing pipelines in scikit-learn	334
Using pipelines in grid searches	336

| Summary | 337 |

Chapter 12: Wrapping Up — 339
- **Approaching a machine learning problem** — 339
- **Building your own estimator** — 341
  - Writing your own OpenCV-based classifier in C++ — 341
  - Writing your own scikit-learn-based classifier in Python — 344
- **Where to go from here?** — 346
- **Summary** — 348

Index — 349

Preface

I'm glad you're here. It's about time we talked about machine learning.

Machine learning is no longer just a buzzword, it is all around us: from protecting your email, to automatically tagging friends in pictures, to predicting what movies you like. As a subfield of data science, machine learning enables computers to learn through experience: to make predictions about the future using collected data from the past.

And the amount of data to be analyzed is enormous! Current estimates put the daily amount of produced data at 2.5 exabytes (or roughly 1 billion gigabytes). Can you believe it? This would be enough data to fill up 10 million blu-ray discs, or amount to 90 years of HD video. In order to deal with this vast amount of data, companies such as Google, Amazon, Microsoft, and Facebook have been heavily investing in the development of data science platforms that allow us to benefit from machine learning wherever we go—scaling from your mobile phone application all the way to supercomputers connected through the cloud.

In other words: this is the time to invest in machine learning. And if it is your wish to become a machine learning practitioner, too—then this book is for you!

But fret not: your application does not need to be as large-scale or influential as the above examples in order to benefit from machine learning. Everyone starts small. Thus, the first step of this book is to introduce you to the essential concepts of statistical learning, such as classification and regression, with the help of simple and intuitive examples. If you have already studied machine learning theory in detail, this book will show you how to put your knowledge into practice. Oh, and don't worry if you are completely new to the field of machine learning—all you need is the willingness to learn.

Once we have covered all the basic concepts, we will start exploring various algorithms such as decision trees, support vector machines, and Bayesian networks, and learn how to combine them with other OpenCV functionality. Along the way, you will learn how to understand the task by understanding the data and how to build fully functioning machine learning pipelines.

As the book progresses, so will your machine learning skills, until you are ready to take on today's hottest topic in the field: deep learning. Combined with the trained skill of knowing how to select the right tool for the task, we will make sure you get comfortable with all relevant machine learning fundamentals.

At the end of the book, you will be ready to take on your own machine learning problems, either by building on the existing source code or developing your own algorithm from scratch!

What this book covers

Chapter 1, *A Taste of Machine Learning*, will gently introduce you to the different subfields of machine learning, and explain how to install OpenCV and other essential tools in the Python Anaconda environment.

Chapter 2, *Working with Data in OpenCV and Python*, will show you what a typical machine learning workflow looks like, and where data comes in to play. I will explain the difference between training and test data, and show you how to load, store, manipulate, and visualize data with OpenCV and Python.

Chapter 3, *First Steps in Supervised Learning*, will introduce you to the topic of supervised learning by reviewing some core concepts, such as classification and regression. You will learn how to implement a simple machine learning algorithm in OpenCV, how to make predictions about the data, and how to evaluate your model.

Chapter 4, *Representing Data and Engineering Features*, will teach you how to get a feel for some common and well-known machine learning datasets and how to extract the interesting stuff from your raw data.

Chapter 5, *Using Decision Trees to Make a Medical Diagnosis*, will show you how to build decision trees in OpenCV, and use them in a variety of classification and regression problems.

Chapter 6, *Detecting Pedestrians with Support Vector Machines*, will explain how to build support vector machines in OpenCV, and how to apply them to detect pedestrians in images.

Chapter 7, *Implementing a Spam Filter with Bayesian Learning*, will introduce you to probability theory, and show you how you can use Bayesian inference to classify emails as spam or not.

Chapter 8, *Discovering Hidden Structures with Unsupervised Learning*, will talk about unsupervised learning algorithms such as k-means clustering and Expectation-Maximization, and show you how they can be used to extract hidden structures in simple, unlabeled datasets.

Chapter 9, *Using Deep Learning to Classify Handwritten Digits*, will introduce you to the exciting field of deep learning. Starting with the perceptron and multi-layer perceptrons, you will learn how to build deep neural networks in order to classify handwritten digits from the extensive MNIST database.

Chapter 10, *Combining Different Algorithms into an Ensemble*, will show you how to effectively combine multiple algorithms into an ensemble in order to overcome the weaknesses of individual learners, resulting in more accurate and reliable predictions.

Chapter 11, *Selecting the Right Model with Hyper-Parameter Tuning*, will introduce you to the concept of model selection, which allows you to compare different machine learning algorithms in order to select the right tool for the task at hand.

Chapter 12, *Wrapping Up*, will conclude the book by giving you some useful tips on how to approach future machine learning problems on your own, and where to find information on more advanced topics.

What you need for this book

You will need a computer, Python Anaconda, and enthusiasm. Lots of enthusiasm. *Why Python?*, you may ask. The answer is simple: it has become the de facto language of data science, thanks to its great number of open source libraries and tools to process and interact with data.

One of these tools is the Python Anaconda distribution, which provides all the scientific computing libraries we could possibly ask for, such as NumPy, SciPy, Matplotlib, Scikit-Learn, and Pandas. In addition, installing OpenCV is essentially a one-liner. No more flipping switches in cc make or compiling from scratch! We will talk about how to install Python Anaconda in Chapter 1, *A Taste of Machine Learning*.

If you have mostly been using OpenCV in combination with C++, that's fine. But, at least for the purpose of this book, I would strongly suggest that you switch to Python. C++ is fine when your task is to develop high-performance code or real-time applications. But when it comes to picking up a new skill, I believe Python to be a fundamentally better choice of language, because you can do more by typing less. Rather than getting annoyed by the syntactic subtleties of C++, or wasting hours trying to convert data from one format into another, Python will help you concentrate on the topic at hand: to become an expert in machine learning.

Who this book is for

Throughout the book, I will assume that you already have a basic knowledge of OpenCV and Python, but that there is always room to learn more.

Conventions

In this book, you will find a number of text styles that distinguish between different kinds of information. Here are some examples of these styles and an explanation of their meaning. Code words in text, database table names, folder names, filenames, file extensions, pathnames, dummy URLs, user input, and Twitter handles are shown as follows: "In Python, we can create a list of integers by using the `list()` command." A block of code is set using the IPython notation, marking user input with `In [X]`, line continuations with ... and corresponding output with `Out[X]`:

```
In [1]: import numpy
   ...: numpy.__version__
Out[1]: '1.11.3'
```

When we wish to draw your attention to a particular part of a code block, the relevant lines or items are set in bold:

```
In [1]: import numpy
   ...: numpy.__version__
Out[1]: '1.11.3'
```

Any command-line input or output is written as follows:

```
$ ipython
```

New terms and **important** words are shown in bold. Words that you see on the screen, for example, in menus or dialog boxes, appear in the text like this: "Clicking the **Next** button moves you to the next screen."

Warnings or important notes appear in a box like this.

Tips and tricks appear like this.

Reader feedback

Feedback from our readers is always welcome. Let us know what you think about this book—what you liked or disliked. Reader feedback is important for us as it helps us develop titles that you will really get the most out of.

To send us general feedback, simply e-mail feedback@packtpub.com, and mention the book's title in the subject of your message.

If there is a topic that you have expertise in and you are interested in either writing or contributing to a book, see our author guide at www.packtpub.com/authors.

Customer support

Now that you are the proud owner of a Packt book, we have a number of things to help you to get the most from your purchase.

Downloading the example code

You can download the latest version of the example code files for this book from GitHub: http://github.com/mbeyeler/opencv-machine-learning. All code is released under the MIT software license, so you are free to use, adapt, and share the code as you see fit. There you will also be able to explore the source code by browsing through the different Jupyter notebooks.

If you get stuck or have questions about the source code, you are welcome to post in our web forum: https://groups.google.com/d/forum/machine-learning-for-opencv. Chances are, someone else has already shared a solution to your specific problem.

Alternatively, you can download the original code files from the date of publication by visiting your account at http://www.packtpub.com. If you purchased this book elsewhere, you can visit http://www.packtpub.com/support and register to have the files e-mailed directly to you.

You can download the code files by following these steps:

1. Log in or register to our website using your e-mail address and password.
2. Hover the mouse pointer on the **SUPPORT** tab at the top.
3. Click on **Code Downloads & Errata**.
4. Enter the name of the book in the **Search** box.
5. Select the book for which you're looking to download the code files.

6. Choose from the drop-down menu where you purchased this book from.
7. Click on **Code Download**.

You can also download the code files by clicking on the **Code Files** button on the book's webpage at the Packt Publishing website. This page can be accessed by entering the book's name in the **Search** box. Please note that you need to be logged in to your Packt account.

Once the file is downloaded, please make sure that you unzip or extract the folder using the latest version of:

- WinRAR / 7-Zip for Windows
- Zipeg / iZip / UnRarX for Mac
- 7-Zip / PeaZip for Linux

The code bundle for the book is also hosted on GitHub at `https://github.com/PacktPublishing/Machine-Learning-For-OpenCV`. We also have other code bundles from our rich catalog of books and videos available at `https://github.com/PacktPublishing/`. Check them out!

Errata

Although we have taken every care to ensure the accuracy of our content, mistakes do happen. If you find a mistake in one of our books—maybe a mistake in the text or the code—we would be grateful if you could report this to us. By doing so, you can save other readers from frustration and help us improve subsequent versions of this book. If you find any errata, please report them by visiting `http://www.packtpub.com/submit-errata`, selecting your book, clicking on the Errata Submission Form link, and entering the details of your errata. Once your errata are verified, your submission will be accepted and the errata will be uploaded to our website or added to any list of existing errata under the Errata section of that title.

To view the previously submitted errata, go to `https://www.packtpub.com/books/content/support` and enter the name of the book in the search field. The required information will appear under the Errata section.

Piracy

Piracy of copyrighted material on the Internet is an ongoing problem across all media. At Packt, we take the protection of our copyright and licenses very seriously. If you come across any illegal copies of our works in any form on the Internet, please provide us with the location address or website name immediately so that we can pursue a remedy.

Please contact us at `copyright@packtpub.com` with a link to the suspected pirated material.

We appreciate your help in protecting our authors and our ability to bring you valuable content.

Questions

If you have a problem with any aspect of this book, you can contact us at `questions@packtpub.com`, and we will do our best to address the problem.

1
A Taste of Machine Learning

So, you have decided to enter the field of **machine learning**. That's great!

Nowadays, machine learning is all around us—from protecting our email, to automatically tagging our friends in pictures, to predicting what movies we like. As a form of **artificial intelligence**, machine learning enables computers to learn through experience: to make predictions about the future using collected data from the past. On top of that, **computer vision** is one of today's most exciting application fields of machine learning, with deep learning and convolutional neural networks driving innovative systems such as self-driving cars and Google's DeepMind.

However, fret not; your application does not need to be as large-scale or world-changing as the previous examples in order to benefit from machine learning. In this chapter, we will talk about why machine learning has become so popular and discuss the kinds of problems that it can solve. We will then introduce the tools that we need in order to solve machine learning problems using OpenCV. Throughout the book, I will assume that you already have a basic knowledge of OpenCV and Python, but that there is always room to learn more.

Are you ready then? Let's go!

Getting started with machine learning

Machine learning has been around for at least 60 years. Growing out of the quest for artificial intelligence, early machine learning systems used hand-coded rules of `if...else` statements to process data and make decisions. Think of a spam filter whose job is to parse incoming emails and move unwanted messages to a spam folder:

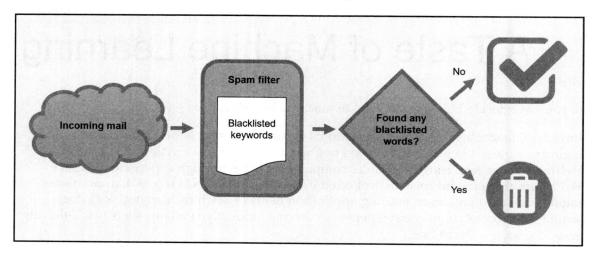

Spam filter

We could come up with a blacklist of words that, whenever they show up in a message, would mark an email as spam. This is a simple example of a hand-coded **expert system**. (We will build a smarter one in `Chapter 7`, *Implementing a Spam Filter with Bayesian Learning*.)

We can think of these expert decision rules to become arbitrarily complicated if we are allowed to combine and nest them in what is known as a **decision tree** (`Chapter 5`, *Using Decision Trees to Make a Medical Diagnosis*). Then, it becomes possible to make more informed decisions that involve a series of decision steps, as shown in the following image:

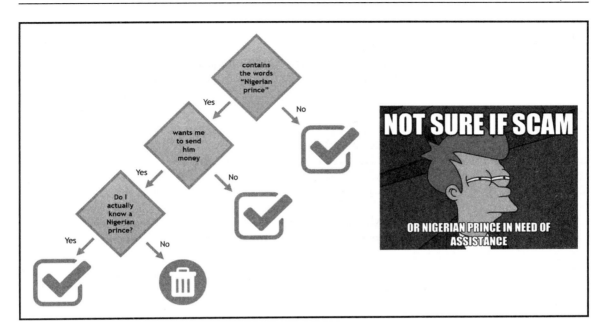

Decision steps in a simple spam filter

Hand-coding these decision rules is sometimes feasible, but has two major disadvantages:

- The logic required to make a decision applies only to a specific task in a single domain. For example, there is no way that we could use this spam filter to tag our friends in a picture. Even if we wanted to change the spam filter to do something slightly different, such as filtering out phishing emails in general, we would have to redesign all the decision rules.
- Designing rules by hand requires a deep understanding of the problem. We would have to know exactly which type of emails constitute spam, including all possible exceptions. This is not as easy as it seems; otherwise, we wouldn't often be double-checking our spam folder for important messages that might have been accidentally filtered out. For other domain problems, it is simply not possible to design the rules by hand.

This is where machine learning comes in. Sometimes, tasks cannot be defined well—except maybe by example—and we would like machines to make sense of and solve the tasks by themselves. Other times, it is possible that, hidden among large piles of data, are important relationships and correlations that we as humans might have missed (see Chapter 8, *Discovering Hidden Structures with Unsupervised Learning*). In these cases, machine learning can often be used to extract these hidden relationships (also known as **data mining**).

A good example of where man-made expert systems have failed is in detecting faces in images. Silly, isn't it? Today, every smart phone can detect a face in an image. However, 20 years ago, this problem was largely unsolved. The reason for this was the way humans think about what constitutes a face was not very helpful to machines. As humans, we tend not to think in pixels. If we were asked to detect a face, we would probably just look for the defining **features** of a face, such as eyes, nose, mouth, and so on. But how would we tell a machine what to look for, when all the machine knows is that images have pixels and pixels have a certain shade of gray? For the longest time, this difference in **image representation** basically made it impossible for a human to come up with a good set of decision rules that would allow a machine to detect a face in an image. We will talk about different approaches to this problem in `Chapter 4`, *Representing Data and Engineering Features*.

However, with the advent of **convolutional neural networks** and **deep learning** (`Chapter 9`, *Using Deep Learning to Classify Handwritten Digits*), machines have become as successful as us when it comes to recognizing faces. All we had to do was simply present a large collection of images of faces to the machine. From there on, the machine was able to discover the set of characteristics that would allow it to identify a face, without having to approach the problem in the same way as we would do. This is the true power of machine learning.

Problems that machine learning can solve

Most machine learning problems belong to one of the following three main categories:

- In **supervised learning**, each data point is labeled or associated with a category or value of interest (`Chapter 3`, *First Steps in Supervised Learning*). An example of a categorical **label** is assigning an image as either a cat or dog. An example of a value label is the sale price associated with a used car. The goal of supervised learning is to study many labeled examples like these (called **training data**) in order to make predictions about future data points (called **test data**). These predictions come in two flavors, such as identifying new photos with the correct animal (called a **classification** problem) or assigning accurate sale prices to other used cars (called a **regression** problem). Don't worry if this seems a little over your head for now—we will have the entirety of the book to nail down the details.
- In **unsupervised learning**, data points have no labels associated with them (`Chapter 8`, *Discovering Hidden Structures with Unsupervised Learning*). Instead, the goal of an unsupervised learning algorithm is to organize the data in some way or to describe its structure. This can mean grouping them into **clusters** or finding different ways of looking at complex data so that they appear simpler.

- In **reinforcement learning**, the algorithm gets to choose an action in response to each data point. It is a common approach in robotics, where the set of sensor readings at one point in time is a data point and the algorithm must choose the robot's next action. It's also a natural fit for **Internet of Things** applications, where the learning algorithm receives a **reward signal** at a short time into the future, indicating how good the decision was. Based on this, the algorithm modifies its strategy in order to achieve the highest reward.

These three main categories are illustrated in the following figure:

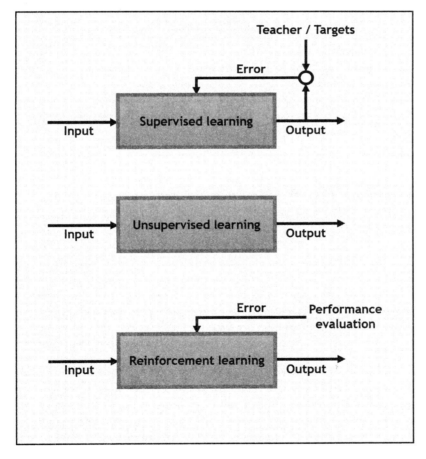

Main machine learning categories

Getting started with Python

Python has become the common language for many data science and machine learning applications, thanks to its great number of open-source libraries for processes such as data loading, data visualization, statistics, image processing, and natural language processing. One of the main advantages of using Python is the ability to interact directly with the code, using a terminal or other tools such as the **Jupyter Notebook**, which we'll look at shortly.

If you have mostly been using OpenCV in combination with C++, I would strongly suggest that you switch to Python, at least for the purpose of studying this book. This decision has not been made out of spite! Quite the contrary: I have done my fair share of C/C++ programming—especially in combination with GPU computing via NVIDIA's **Compute Unified Device Architecture** (**CUDA**)—and like it a lot. However, I consider Python to be a better choice fundamentally if you want to pick up a new topical skill, because you can do more by typing less. This will help reduce the **cognitive load**. Rather than getting annoyed by the syntactic subtleties of C++ or wasting hours trying to convert data from one format to another, Python will help you concentrate on the topic at hand: becoming an expert in machine learning.

Getting started with OpenCV

Being the avid user of OpenCV that I believe you are, I probably don't have to convince you about the power of OpenCV.

Built to provide a common infrastructure for computer vision applications, OpenCV has become a comprehensive set of both classic and state-of-the-art computer vision and machine learning algorithms. According to their own documentation, OpenCV has a user community of more than 47,000 people and has been downloaded over seven million times. That's pretty impressive! As an open-source project, it is very easy for researchers, businesses, and government bodies to utilize and modify already available code.

This being said, a number of open-source machine learning libraries have popped up since the recent machine learning boom that provide far more functionality than OpenCV. A prominent example is **scikit-learn**, which provides a number of state-of-the-art machine learning algorithms as well as a wealth of online tutorials and code snippets. As OpenCV was developed mainly to provide computer vision algorithms, its machine learning functionality is restricted to a single module, called `ml`. As we will see in this book, OpenCV still provides a number of state-of-the-art algorithms, but sometimes lacks a bit in functionality. In these rare cases, instead of reinventing the wheel, we will simply use scikit-learn for our purposes.

Last but not least, installing OpenCV using the Python Anaconda distribution is essentially a one-liner!

If you are a more advanced user who wants to build real-time applications, OpenCV's algorithms are well-optimized for this task, and Python provides several ways to speed up computations where it is necessary (using, for example, **Cython** or parallel processing libraries such as **joblib** or **dask**).

Installation

Before we get started, let's make sure that we have all the tools and libraries installed that are necessary to create a fully functioning data science environment. After downloading the latest code for this book from GitHub, we are going to install the following software:

- Python's Anaconda distribution, based on Python 3.5 or higher
- OpenCV 3.1 or higher
- Some supporting packages

Don't feel like installing stuff? You can also visit http://beta.mybinder.org/v2/gh/mbeyeler/opencv-machine-learning/master, where you will find all the code for this book in an interactive, executable environment and 100% free and open source, thanks to the **Binder** project.

Getting the latest code for this book

You can get the latest code for this book from GitHub, https://github.com/mbeyeler/opencv-machine-learning. You can either download a .zip package (beginners) or clone the repository using git (intermediate users).

Git is a **version control system** that allows you to track changes in files and collaborate with others on your code. In addition, the web platform **GitHub.com** makes it easy for me to share my code with you on a public server. As I make improvements to the code, you can easily update your local copy, file bug reports, or suggest code changes.

If you choose to go with git, the first step is to make sure it is installed (https://git-scm.com/downloads).

Then, open a **terminal** (or **command prompt**, as it is called in Windows):

- On Windows 10, right-click on the Start Menu button, and select **Command Prompt**.
- On Mac OS X, press *Cmd* + *Space* to open spotlight search, then type `terminal`, and hit *Enter*.
- On Ubuntu and friends, press *Ctrl* + *Alt* + *T*. On Red Hat, right-click on the desktop and choose **Open Terminal** from the menu.

Navigate to a directory where you want the code downloaded, for example:

```
$ cd Desktop
```

Then you can grab a local copy of the latest code by typing the following:

```
$ git clone https://github.com/mbeyeler/opencv-machine-learning.git
```

This will download the latest code in a folder called `opencv-machine-learning`.

After a while, the code might change online. In that case, you can update your local copy by running the following command from within the `opencv-machine-learning` directory:

```
$ git pull origin master
```

Getting to grips with Python's Anaconda distribution

Anaconda is a free Python distribution developed by Continuum Analytics that is made for scientific computing. It works across Windows, Linux, and Mac OS X platforms and is free even for commercial use. However, the best thing about it is that it comes with a number of preinstalled packages that are essential for data science, math, and engineering. These packages include the following:

- **NumPy**: A fundamental package for scientific computing in Python, which provides functionality for multidimensional arrays, high-level mathematical functions, and pseudo-random number generators
- **SciPy**: A collection of functions for scientific computing in Python, which provides advanced linear algebra routines, mathematical function optimization, signal processing, and so on

- **scikit-learn**: An open-source machine learning library in Python, which provides useful helper functions and infrastructure that OpenCV lacks
- **Matplotlib**: The primary scientific plotting library in Python, which provides functionality for producing line charts, histograms, scatter plots, and so on
- **Jupyter Notebook**: An interactive environment for the running of code in a web browser

An installer for our platform of choice (Windows, Mac OS X, or Linux) can be found on the Continuum website, https://www.continuum.io/Downloads. I recommend using the Python 3.6-based distribution, as Python 2 is no longer under active development.

To run the installer, do one of the following:

- On Windows, double-click on the .exe file and follow the instructions on the screen
- On Mac OS X, double-click on the .pkg file and follow the instructions on the screen
- On Linux, open a terminal and run the .sh script using bash:

```
$ bash Anaconda3-4.3.0-Linux-x86_64.sh   # Python 3.6 based
$ bash Anaconda2-4.3.0-Linux-x64_64.sh   # Python 2.7 based
```

In addition, Python Anaconda comes with conda—a simple package manager similar to apt-get on Linux. After successful installation, we can install new packages in the terminal using the following command:

```
$ conda install package_name
```

Here, package_name is the actual name of the package that we want to install.

Existing packages can be updated using the following command:

```
$ conda update package_name
```

We can also search for packages using the following command:

```
$ anaconda search -t conda package_name
```

This will bring up a whole list of packages available through individual users. For example, searching for a package named `opencv`, we get the following hits:

```
C:\Users\mbeyeler>anaconda search -t conda opencv
Using Anaconda API: https://api.anaconda.org
Run 'anaconda show <USER/PACKAGE>' to get more details:
Packages:
     Name                         | Version    | Package Types   | Platforms
     -------------------------    | ------     | --------------- | ---------------
     ???/opencv                   | 2.4.7      | conda           | win-64
                                                : http://opencv.org/
     Changxu/opencv3              | 3.1.0_dev  | conda           | linux-64
     Definiter/opencv             | 2.4.12     | conda           | linux-64
     FlyEM/opencv                 | 2.4.10.1   | conda           | linux-64, osx-64
                                                : Open source computer vision C++ library
     JaimeIvanCervantes/opencv    | 2.4.9.99   | conda           | linux-64
     RahulJain/opencv             | 2.4.12     | conda           | linux-64, win-64, osx-64
     anaconda-backup/opencv       | 3.1.0      | conda           | linux-64, linux-32, osx-64
     anaconda/opencv              | 3.1.0      | conda           | linux-64, linux-32, osx-64
     andywocky/opencv             | 2.4.9      | conda           | osx-64
     asmeurer/opencv              | 2.4.9      | conda           | osx-64
     bgreen-litl/opencv           | 2.4.9      | conda           | osx-64
     bwsprague/opencv             | 2.4.9.1    | conda           | osx-64
     cfobel/opencv-helpers        | 0.1.post1  | conda           | win-32
     clg_boar/opencv3             | 3.0.0      | conda           | linux-64, win-64
     clinicalgraphics/opencv      |            | conda           | linux-64, win-32, win-64, linux-32, osx-64
     conda-forge/opencv           | 3.1.0      | conda           | linux-64, win-32, win-64, osx-64
                                                : Computer vision and machine learning software library.
```

Searching for OpenCV packages provided by different conda users.

This will bring up a long list of users who have OpenCV packages installed, where we can locate users that have our version of the software installed on our own platform. A package called `package_name` from a user called `user_name` can then be installed as follows:

```
$ conda install -c user_name package_name
```

Finally, `conda` provides something called an **environment**, which allows us to manage different versions of Python and/or packages installed in them. This means we could have one environment where we have all packages necessary to run OpenCV 2.4 with Python 2.7, and another where we run OpenCV 3.2 with Python 3.6. In the following section, we will create an environment that contains all the packages needed to run the code in this book.

Installing OpenCV in a conda environment

In a terminal, navigate to the directory where you downloaded the code:

```
$ cd Desktop/opencv-machine-learning
```

Before we create a new conda environment, we want to make sure we added the Conda-Forge channel to our list of trusted conda channels:

```
$ conda config --add channels conda-forge
```

The Conda-Forge channel is led by an open-source community that provides a wide variety of code recipes and software packages (for more info, see https://conda-forge.github.io). Specifically, it provides an OpenCV package for 64-bit Windows, which will simplify the remaining steps of the installation.

Then run the following command to create a conda environment based on Python 3.5, which will also install all the necessary packages listed in the file requirements.txt in one fell swoop:

```
$ conda create -n Python3 python=3.5 --file requirements.txt
```

To activate the environment, type one of the following, depending on your platform:

```
$ source activate Python3    # on Linux / Mac OS X
$ activate Python3            # on Windows
```

Once we close the terminal, the session will be deactivated—so we will have to run this last command again the next time we open a terminal. We can also deactivate the environment by hand:

```
$ source deactivate   # on Linux / Mac OS X
$ deactivate          # on Windows
```

And done!

Verifying the installation

It's a good idea to double-check our installation. While our terminal is still open, we fire up **IPython**, which is an interactive shell to run Python commands:

```
$ ipython
```

Now make sure that you are running (at least) Python 3.5 and not Python 2.7. You might see the version number displayed in IPython's welcome message. If not, you can run the following commands:

```
In [1]: import sys
   ...      print(sys.version)
            3.5.3 |Continuum Analytics, Inc.| (default, Feb 22 2017,
21:28:42) [MSC v.1900 64 bit (AMD64)]
```

Now try to import OpenCV:

```
In [2]: import cv2
```

You should get no error messages. Then, try to find out the version number:

```
In [3]: cv2.__version__
Out[3]: '3.1.0'
```

Make sure that the Python version reads 3.5 or 3.6, but not 2.7. Additionally, make sure that OpenCV's version number reads at least **3.1.0**; otherwise, you will not be able to use some OpenCV functionality later on.

OpenCV 3 is actually called `cv2`. I know it's confusing. Apparently, the reason for this is that the **2** does not stand for the version number. Instead, it is meant to highlight the difference between the underlying C API (which is denoted by the `cv` prefix) and the C++ API (which is denoted by the `cv2` prefix).

You can then exit the IPython shell by typing **exit** - or hitting *Ctrl + D* and confirming that you want to quit.

Alternatively, you can run the code in a web browser thanks to **Jupyter Notebook**. If you have never heard of Jupyter Notebooks or played with them before, trust me - you will love them! If you followed the directions as mentioned earlier and installed the Python Anaconda stack, Jupyter is already installed and ready to go. In a terminal, type as follows:

```
$ jupyter notebook
```

This will automatically open a browser window, showing a list of files in the current directory. Click on the `opencv-machine-learning` folder, then on the `notebooks` folder, and voila! Here you will find all the code for this book, ready for you to be explored:

Chapter 1

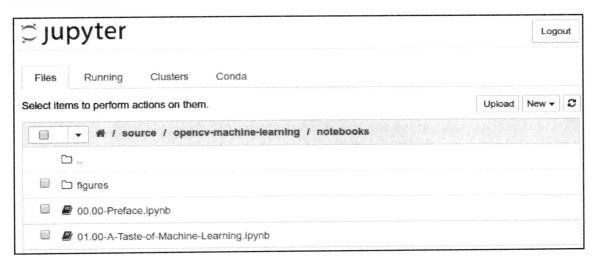

Beginning of the list of Jupyter Notebooks that come with this book

The notebooks are arranged by chapter and section. For the most part, they contain only the relevant code, but no additional information or explanations. These are reserved for those who support our effort by buying this book - so thank you!

Simply click on a notebook of your choice, such as 01.00-A-Taste-of-Machine-Learning.ipynb, and you will be able to run the code yourself by selecting *Kernel > Restart & Run All*:

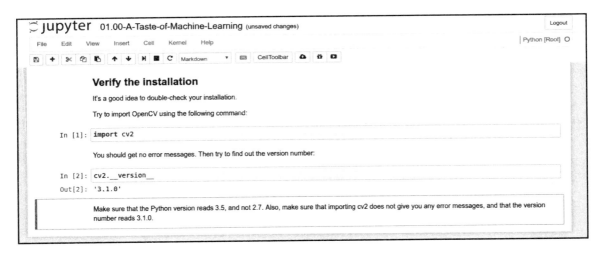

Example excerpt of this chapter's Jupyter Notebook

There are a few handy keyboard shortcuts for navigating Jupyter Notebooks. However, the only ones that you need to know about right now are the following:

1. Click in a cell in order to edit it.
2. While the cell is selected, hit *Ctrl + Enter* to execute the code in it.
3. Alternatively, hit *Shift + Enter* to execute a cell and select the cell below it.
4. Hit *Esc* to exit write mode, then hit *A* to insert a cell above the currently selected one and *B* to insert a cell below.

Check out all the keyboard shortcuts by clicking on *Help > Keyboard Shortcut*, or take a quick tour by clicking on *Help > User Interface Tour*.

However, I strongly encourage you to follow along the book by actually typing out the commands yourself, preferably in an IPython shell or an empty Jupyter Notebook. There is no better way to learn how to code than by getting your hands dirty. Even better if you make mistakes—we have all been there. At the end of the day, it's all about *learning by doing*!

Getting a glimpse of OpenCV's ML module

Starting with OpenCV 3.1, all machine learning-related functions in OpenCV have been grouped into the `ml` module. This has been the case for the C++ API for quite some time. You can get a glimpse of what's to come by displaying all functions in the `ml` module:

```
In [4]: dir(cv2.ml)
Out[4]: ['ANN_MLP_BACKPROP',
         'ANN_MLP_GAUSSIAN',
         'ANN_MLP_IDENTITY',
         'ANN_MLP_NO_INPUT_SCALE',
         'ANN_MLP_NO_OUTPUT_SCALE',
         ...
         '__spec__']
```

If you have installed an older version of OpenCV, the `ml` module might not be present. For example, the k-nearest neighbor algorithm (which we will talk about in Chapter 3, *First Steps in Supervised Learning*) used to be called `cv2.KNearest()` but is now called `cv2.ml.KNearest_create()`. In order to avoid confusion throughout the book, I therefore recommend using at least OpenCV 3.1.

Summary

In this chapter, we talked about machine learning at a high abstraction level: what it is, why it is important, and what kinds of problems it can solve. We learned that machine learning problems come in three flavors: supervised learning, unsupervised learning, and reinforcement learning. We talked about the prominence of supervised learning, and that this field can be further divided into two subfields: classification and regression. Classification models allow us to categorize objects into known classes (such as animals into cats and dogs), whereas regression analysis can be used to predict continuous outcomes of target variables (such as the sales price of used cars).

We also learned how to set up a data science environment using the Python Anaconda distribution, how to get the latest code of this book from GitHub, and how to run code in a Jupyter Notebook.

With these tools in hand, we are now ready to start talking about machine learning in more detail. In the next chapter, we will look at the inner workings of machine learning systems and learn how to work with data in OpenCV with the help of common Pythonic tools such as NumPy and Matplotlib.

2
Working with Data in OpenCV and Python

Now that we have whetted our appetite for machine learning, it is time to delve a little deeper into the different parts that make up a typical machine learning system.

Far too often, you hear someone throw around the phrase, *just apply machine learning to your data!*, as if that will instantly solve all your problems. You can imagine that the reality of this is much more intricate. Although, I will admit that nowadays it is incredibly easy to build your own machine learning system simply by cutting and pasting just a few lines of code from the internet. However, in order to build a system that is truly powerful and effective, it is essential to have a firm grasp of the underlying concepts and an intimate knowledge of the strengths and weaknesses of each method. So don't worry if you aren't considering yourself a machine learning expert just yet. Good things take time.

Earlier, I described machine learning as a subfield of artificial intelligence. This might be true—mainly for historical reasons—but most often, machine learning is simply about **making sense of data**. Therefore, it might be more suitable to think of machine learning as a subfield of **data science**, where we build mathematical models to help understand data.

Hence, this chapter is all about data. We want to learn how data fits in with machine learning, and how to work with data using the tools of our choice: OpenCV and Python.

Specifically, we want to address the following questions:

- What does a typical machine learning workflow look like; where does data come into play?
- What are training data and test data; what are they good for?
- How do I load, store, edit, and visualize data with OpenCV and Python?

Understanding the machine learning workflow

As mentioned earlier, machine learning is all about building mathematical models in order to understand data. The learning aspect enters this process when we give a machine learning model the capability to adjust its **internal parameters**; we can tweak these parameters so that the model explains the data better . In a sense, this can be understood as the model learning from the data. Once the model has learned enough—whatever that means—we can ask it to explain newly observed data.

This process is illustrated in the following figure:

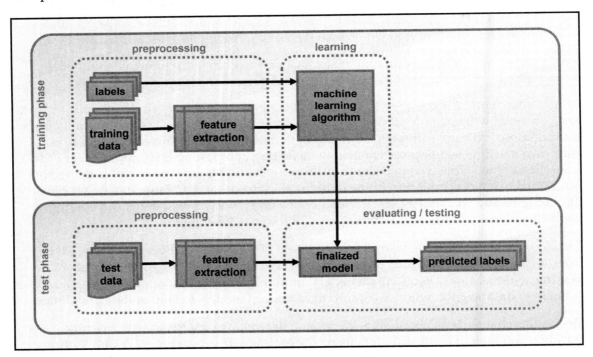

A typical workflow to tackle machine learning problems

Let's break it down step by step.

The first thing to notice is that machine learning problems are always split into (at least) two distinct phases:

- A **training phase**, during which we aim to train a machine learning model on a set of data that we call the **training dataset**
- A **test phase**, during which we evaluate the learned (or finalized) machine learning model on a new set of never-before-seen data that we call the **test dataset**

The importance of splitting our data into a training set and test set cannot be understated. We always evaluate our models on an independent test set because we are interested in knowing how well our models **generalize to new data**. In the end, isn't this what learning is all about—be it machine learning or human learning? Think back to school when you were a learner yourself: the problems you had to solve as part of your homework would never show up in exactly the same form in the final exam. The same scrutiny should be applied to a machine learning model; we are not so much interested in how well our models can memorize a set of data points (such as a homework problem), but we want to know how our models will use what they have learned to solve new problems (such as the ones that show up in a final exam) and explain new data points.

The workflow of an advanced machine learning problem will typically include a third set of data termed **validation dataset**. For now, this distinction is not important. A validation set is typically formed by further partitioning the training dataset. It is used in advanced concepts such as model selection, which we will talk about in Chapter 11, *Selecting the Right Model with Hyper-Parameter Tuning*, when we have become proficient in building machine learning systems.

The next thing to notice is that machine learning is really all about the **data**. Data enters the previously described workflow diagram in its raw form—whatever that means—and is used in both training and test phases. Data can be anything from images and movies to text documents and audio files. Therefore, in its raw form, data might be made of pixels, letters, words, or even worse: pure bits. It is easy to see that data in such a raw form might not be very convenient to work with. Instead, we have to find ways to **preprocess** the data in order to bring it into a form that is easy to **parse**.

Data preprocessing comes in two stages:

- **Feature selection**: This is the process of identifying important attributes (or features) in the data. Possible features of an image might be the location of edges, corners, or ridges. You might already be familiar with some more advanced feature descriptors that OpenCV provides, such as **speeded up robust features (SURF)** or the **histogram of oriented gradients (HOG)**. Although these features can be applied to any image, they might not be that important (or work that well) for our specific task. For example, if our task was to distinguish between clean and dirty water, the most important feature might turn out to be the color of the water, and the use of SURF or HOG features might not help us much.
- **Feature extraction**: This is the actual process of transforming the raw data into the desired **feature space**. An example would be the **Harris operator**, which allows us to extract corners (that is, a selected feature) in an image.

A more advanced topic is the process of inventing informative features, which is known as **feature engineering**. After all, before it was possible for people to select from popular features, someone had to invent them first. This is often more important for the success of our algorithm than the choice of the algorithm itself. We will talk about feature engineering extensively in `Chapter 4`, *Representing Data and Engineering Features*.

> Don't let naming **conventions** confuse you! Sometimes feature selection and feature extraction are hard to distinguish, mainly because of how stuff is named. For example, SURF stands for both the feature extractor as well as the actual name of the features. The same is true for the **scale-invariant feature transform (SIFT)**, which is a feature extractor that yields what is known as **SIFT features**.

A last point to be made is that in supervised learning, every data point must have a **label**. A label identifies a data point of either belonging to a certain class of things (such as cat or dog) or of having a certain value (such as the price of a house). At the end of the day, the goal of a supervised machine learning system is to predict the label of all data points in the test set (as shown in the previous figure). We do this by learning regularities in the training data, using the labels that come with it, and then testing our performance on the test set.

Therefore, in order to build a functioning machine learning system, we first have to cover how to load, store, and manipulate data. How do you even do that in OpenCV with Python?

Dealing with data using OpenCV and Python

Although raw data can come from a variety of sources and in a wide range of formats, it will help us to think of all data fundamentally as **arrays of numbers**. For example, images can be thought of as simply 2D arrays of numbers representing pixel brightness across an area. Sound clips can be thought of 1D arrays of intensity over time. For this reason, efficient storage and manipulation of numerical arrays is absolutely fundamental to machine learning.

If you have mostly been using OpenCV's C++ **application programming interface (API)** and plan on continuing to do so, you might find that dealing with data in C++ can be a bit of a pain. Not only will you have to deal with the syntactic overhead of the C++ language, but you will also have to wrestle with different data types and cross-platform compatibility issues.

This process is radically simplified if you use OpenCV's Python API because you automatically get access to a large number of open-source packages from the **Scientific Python (SciPy)** community. Case in point is the **Numerical Python (NumPy)** package, around which most scientific computing tools are built.

Starting a new IPython or Jupyter session

Before we can get our hands on NumPy, we need to open an IPython shell or start a Jupyter Notebook:

1. Open a terminal like we did in the previous chapter, and navigate to the `opencv-machine-learning` directory:

   ```
   $ cd Desktop/opencv-machine-learning
   ```

2. Activate the conda environment we created in the previous chapter:

   ```
   $ source activate Python3    # Mac OS X / Linux
   $ activate Python3           # Windows
   ```

3. Start a new IPython or Jupyter session:

   ```
   $ ipython              # for an IPython session
   $ jupyter notebook     # for a Jupyter session
   ```

Working with Data in OpenCV and Python

If you chose to start an IPython session, the program should have greeted you with a welcome message such as the following:

```
$ ipython
Python 3.5.2 |Continuum Analytics, Inc.| (default, Jul 5 2016, 11:41:13)
[MSC v.1900 64 bit (AMD64)]
Type "copyright", "credits" or "license" for more information.

IPython 3.5.0 -- An enhanced Interactive Python.
?         -> Introduction and overview of IPython's features.
%quickref -> Quick reference.
help      -> Python's own help system.
object?   -> Details about 'object', use 'object??' for extra details.

In [1]:
```

The line starting with `In [1]` is where you type in your regular Python commands. In addition, you can also use the *Tab* key while typing the names of variables and functions in order to have IPython automatically complete them.

A limited number of system shell commands work, too—such as `ls` and `pwd`. You can run any shell command by prefixing it with a !, such as `!ping www.github.com`. For more information, check out the official IPython reference at https://ipython.org/ipython-doc/3/interactive/tutorial.html.

If you chose to start a Jupyter session, a new window should have opened in your web browser that is pointing to `http://localhost:8888`. You want to create a new notebook by clicking on **New** in the top-right corner and selecting **Notebooks (Python3)**:

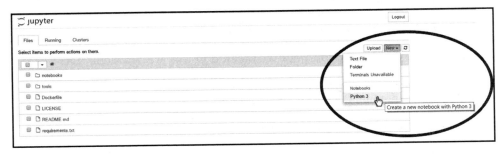

Creating a new Jupyter Notebook

This will open a new window that looks like this:

A new Jupyter Notebook

The cell labeled with In [ ] is the same as the command line in an IPython session. Now you can start typing your Python code!

Dealing with data using Python's NumPy package

If you followed the advice outlined in the previous chapter and installed the Python Anaconda stack, you already have NumPy installed and are ready to go. If you are of the do-it-yourself type, you can go to http://www.numpy.org and follow the installation instructions found there.

I informed you earlier that it would be okay if you weren't a Python expert yet. Who knows, perhaps you're just now switching from OpenCV's C++ API. This is all fine. I wanted to give you a quick overview on how to get started with NumPy. If you are a more advanced Python user, you may simply skip this subsection.

Once you are familiar with NumPy, you will find that most scientific computing tools in the Python world are built around it. This includes OpenCV, so the time spent on learning NumPy is guaranteed to benefit you in the end.

Importing NumPy

Once you start a new IPython or Jupyter session, you can import the NumPy module and verify its version as follows:

```
In [1]: import numpy
In [2]: numpy.__version__
Out[2]: '1.11.3'
```

Recall that in the Jupyter Notebook you can hit *Ctrl + Enter* to execute a cell once you have typed the command. Alternatively, *Shift + Enter* executes the cell and automatically inserts or selects the cell below it. Check out all the keyboard shortcuts by clicking on *Help > Keyboard Shortcut*, or take a quick tour by clicking on *Help > User Interface Tour*.

For the pieces of the package discussed here, I would recommend using NumPy version 1.8 or later. By convention, you'll find that most people in the scientific Python world will import NumPy using `np` as an alias:

```
In [3]: import numpy as np
In [4]: np.__version__
Out[4]: '1.11.3'
```

Throughout this chapter, and the rest of the book, we will stick to the same convention.

Understanding NumPy arrays

You might already know that Python is a **weakly-typed language**. This means that you do not have to specify a data type whenever you create a new variable. For example, the following will automatically be represented as an integer:

```
In [5]: a = 5
```

You can double-check this by typing as follows:

```
In [6]: type(a)
Out[6]: int
```

As the standard Python implementation is written in C, every Python object is basically a C structure in disguise. This is true even for integers in Python, which are actually pointers to compound C structures that contain more than just the **raw** integer value. Therefore, the default C data type used to represent Python integers will depend on your system architecture (that is, whether it is a 32-bit or 64-bit platform).

Going a step further, we can create a list of integers using the `list()` command, which is the standard multielement container in Python. The `range(x)` function will spell out all integers from 0 up to `x-1`:

```
In [7]: int_list = list(range(10))
   ...  int_list
Out[7]: [0, 1, 2, 3, 4, 5, 6, 7, 8, 9]
```

Similarly, we can create a list of strings by telling Python to iterate over all the elements in the integer list, `int_list`, and applying the `str()` function to each element:

```
In [8]: str_list = [str(i) for i in int_list]
   ...: str_list
Out[8]: ['0', '1', '2', '3', '4', '5', '6', '7', '8', '9']
```

However, lists are not very flexible to do math on. Let's say, for example, we wanted to multiply every element in `int_list` by a factor of 2. A naive approach might be to do the following—but see what happens to the output:

```
In [9]: int_list * 2
Out[9]: [0, 1, 2, 3, 4, 5, 6, 7, 8, 9, 0, 1, 2, 3, 4, 5, 6, 7, 8, 9]
```

Python created a list whose content is simply all elements of `int_list` produced twice; this is not what we wanted!

This is where NumPy comes in. NumPy has been designed specifically to make array arithmetic in Python easy. We can quickly convert the list of integers into a NumPy array:

```
In [10]: import numpy as np
    ...: int_arr = np.array(int_list)
    ...: int_arr
Out[10]: array([0, 1, 2, 3, 4, 5, 6, 7, 8, 9])
```

Let's see what happens now when we try to multiply every element in the array:

```
In [11]: int_arr * 2
Out[11]: array([ 0,  2,  4,  6,  8, 10, 12, 14, 16, 18])
```

Now we got it right! The same works with addition, subtraction, division, and many other functions.

In addition, every NumPy array comes with the following attributes:

- `ndim`: The number of dimensions
- `shape`: The size of each dimension
- `size`: The total number of elements in the array
- `dtype`: The data type of the array (for example, `int`, `float`, `string`, and so on)

Let's check these preceding attributes for our integer array:

```
In [12]: print("int_arr ndim: ", int_arr.ndim)
    ...: print("int_arr shape: ", int_arr.shape)
    ...: print("int_arr size: ", int_arr.size)
    ...: print("int_arr dtype: ", int_arr.dtype)
```

```
Out[12]: int_arr ndim: 1
...      int_arr shape: (10,)
...      int_arr size: 10
...      int_arr dtype: int64
```

From these outputs, we can see that our array contains only one dimension, which contains ten elements, and all elements are 64-bit integers. Of course, if you are executing this code on a 32-bit machine, you might find `dtype: int32`.

Accessing single array elements by indexing

If you are familiar with Python's standard list indexing, indexing in NumPy will feel quite familiar. In a 1D array, the *i*-th value (counting from zero) can be accessed by specifying the desired index in square brackets, just as with Python lists:

```
In [13]: int_arr
Out[13]: array([0, 1, 2, 3, 4, 5, 6, 7, 8, 9])
In [14]: int_arr[0]
Out[14]: 0
In [15]: int_arr[3]
Out[15]: int_arr[3]
```

To index from the end of the array, you can use negative indices:

```
In [16]: int_arr[-1]
Out[16]: 9
In [17]: int_arr[-2]
Out[17]: 8
```

There are a few other cool tricks for **slicing arrays**, as follows:

```
In [18]: int_arr[2:5]:   # from index 2 up to index 5 - 1
Out[18]: array([2, 3, 4])
In [19]: int_arr[:5]     # from the beginning up to index 5 - 1
Out[19]: array([0, 1, 2, 3, 4])
In [20]: int_arr[5:]     # from index 5 up to the end of the array
Out[20]: array([5, 6, 7, 8, 9])
In [21]: int_arr[::2]    # every other element
Out[21]: array([0, 2, 4, 6, 8])
In [22]: int_arr[::-1]   # the entire array in reverse order
Out[22]: array([0, 1, 2, 3, 4, 5, 6, 7, 8, 9])
```

I encourage you to play around with these arrays yourself!

The general form of slicing arrays in NumPy is the same as it is for standard Python lists. In order to access a slice of an array x, use `x[start:stop:step]`. If any of these are unspecified, they default to the values start=0, stop=*size of dimension*, step=1.

Creating multidimensional arrays

Arrays need not be limited to lists. In fact, they can have an arbitrary number of dimensions. In machine learning, we will often deal with at least 2D arrays, where the column index stands for the values of a particular feature and the rows contain the actual feature values.

With NumPy, it is easy to create multidimensional arrays from scratch. Let's say that we want to create an array with three rows and five columns, with all the elements initialized to zero. If we don't specify a data type, NumPy will default to using floats:

```
In [23]: arr_2d = np.zeros((3, 5))
    ...  arr_2d
Out[23]: array([[ 0.,  0.,  0.,  0.,  0.],
    ...         [ 0.,  0.,  0.,  0.,  0.],
    ...         [ 0.,  0.,  0.,  0.,  0.]])
```

As you probably know from your OpenCV days, this could be interpreted as a 3 x 5 grayscale image with all pixels set to 0 (black). Analogously, if we wanted to create a tiny 2 x 4 pixel image with three color channels (R, G, B), but all pixels set to white, we would use NumPy to create a 3D array with the dimensions, 3 x 2 x 4:

```
In [24]: arr_float_3d = np.ones((3, 2, 4))
    ...  arr_float_3d
Out[24]: array([[[ 1.,  1.,  1.,  1.],
    ...          [ 1.,  1.,  1.,  1.]],
    ...
    ...         [[ 1.,  1.,  1.,  1.],
    ...          [ 1.,  1.,  1.,  1.]],
    ...
    ...         [[ 1.,  1.,  1.,  1.],
    ...          [ 1.,  1.,  1.,  1.]]])
```

Here, the first dimension defines the color channel (red, green, blue, green, and red in OpenCV). Thus, if this was real image data, we could easily grab the color information in the first channel by slicing the array:

```
In [25]: arr_float_3d[0, :, :]
Out[25]: array([[ 1.,  1.,  1.,  1.],
    ...         [ 1.,  1.,  1.,  1.]])
```

In OpenCV, images either come as 32-bit float arrays with values between 0 and 1 or they come as 8-bit integer arrays with values between 0 and 255. Hence, we can also create a 2 x 4 pixel, all-white RGB image using 8-bit integers by specifying the `dtype` attribute of the NumPy array and multiplying all the ones in the array by 255:

```
In [26]: arr_uint_3d = np.ones((3, 2, 4), dtype=np.uint8) * 255
    ...  arr_unit_3d
Out[26]: array([[[255, 255, 255, 255],
    ...          [255, 255, 255, 255]],
    ...
    ...         [[255, 255, 255, 255],
    ...          [255, 255, 255, 255]],
    ...
    ...         [[255, 255, 255, 255],
    ...          [255, 255, 255, 255]]], dtype=uint8)
```

We will look at more advanced array manipulations in later chapters.

Loading external datasets in Python

Thanks to the SciPy community, there are tons of resources out there for getting our hands on some data.

A particularly useful resource comes in the form of the `sklearn.datasets` package of **scikit-learn**. This package comes preinstalled with some small datasets that do not require us to download any files from external websites. These datasets include the following:

- `load_boston`: The Boston dataset contains housing prices in different suburbs of Boston along with a number of interesting features such as per capita crime rate by town, proportion of residential land, non-retail business, and so on
- `load_iris`: The Iris dataset contains three different types of iris flowers (setosa, versicolor, and virginica), along with four features describing the width and length of the sepals and petals
- `load_diabetes`: The diabetes dataset lets us classify patients as having diabetes or not, based on features such as patient age, sex, body mass index, average blood pressure, and six blood serum measurements
- `load_digits`: The digits dataset contains 8 x 8 pixel images of digits 0-9
- `load_linnerud`: The Linnerud dataset contains three physiological and three exercise variables measured on twenty middle-aged men in a fitness club

In addition, scikit-learn allows us to download datasets directly from external repositories, such as the following:

- `fetch_olivetti_faces`: The Olivetta face dataset contains ten different images each of 40 distinct subjects
- `fetch_20newsgroups`: The 20 newsgroup dataset contains around 18,000 newsgroups posts on 20 topics

Even better, it is possible to download datasets directly from the machine learning database at http://mldata.org. For example, to download the MNIST dataset of handwritten digits, simply type as follows:

```
In [1]: from sklearn import datasets
In [2]: mnist = datasets.fetch_mldata('MNIST original')
```

Note that this might take a while, depending on your internet connection. The MNIST database contains a total of 70,000 examples of handwritten digits (28 x 28 pixel images, labeled from 0 to 9). Data and labels are delivered in two separate containers, which we can inspect as follows:

```
In [3]: mnist.data.shape
Out[3]: (70000, 784)
In [4]: mnist.target.shape
Out[4]: (70000,)
```

Here, we can see that `mnist.data` contains 70,000 images of 28 x 28 = 784 pixels each. Labels are stored in `mnist.target`, where there is only one label per image.

We can further inspect the values of all targets, but we don't just want to print them all. Instead, we are interested to see all distinct target values, which is easy to do with NumPy:

```
In [5]: np.unique(mnist.target)
Out[5]: array([0., 1., 2., 3., 4., 5., 6., 7., 8., 9.])
```

Another Python library for data analysis that you should have heard about is **Pandas** (http://pandas.pydata.org). Pandas implements a number of powerful data operations for both databases and spreadsheets. However great the library, at this point, Pandas is a bit too advanced for our purposes.

Visualizing the data using Matplotlib

Knowing how to load data is of limited use if we don't know how to look at the data. Thankfully, there is **Matplotlib**!

Matplotlib is a multiplatform data visualization library built on NumPy arrays—see, I promised you NumPy would show up again. It was conceived by John Hunter in 2002, originally designed as a patch to IPython to enable interactive MATLAB-style plotting from the command line. In more recent years, newer and shinier tools have popped up to eventually replace Matplotlib (such as `ggplot` and `ggvis` in the R language), but Matplotlib remains essential as a well-tested, cross-platform graphics engine.

Importing Matplotlib

You might be in luck again: if you followed the advice outlined in the previous chapter and installed the Python Anaconda stack, you already have Matplotlib installed and are ready to go. Otherwise, you might want to visit http://matplotlib.org for installation instructions.

Just as we used `np` shorthand for NumPy, we will use some standard shorthand for the Matplotlib imports:

```
In [1]: import matplotlib as mpl
In [2]: import matplotlib.pyplot as plt
```

The `plt` interface is what we will use most often, as we shall see throughout the book.

Producing a simple plot

Without further ado, let's create our first plot.

Let's say that we want to produce a simple line plot of the sine function, `sin(x)`. We want the function to be evaluated at all points on the *x*-axis where $0 \leq x < 10$. We will use NumPy's `linspace` function to create a linear spacing on the x axis, from x values 0 to 10, and a total of 100 sampling points:

```
In [3]: import numpy as np
In [4]: x = np.linspace(0, 10, 100)
```

We can evaluate the `sin` function at all points x using NumPy's `sin` function and visualize the result by calling plt's `plot` function:

```
In [5]: plt.plot(x, np.sin(x))
```

Did you try it yourself? What happened? Did anything show up?

The thing is, depending on where you are running this script, you might not see anything. There are the following possibilities to consider:

- Plotting from a .py script: If you are running Matplotlib from a script, you need to simply call as follows:

    ```
    plt.show()
    ```

    Call this at the end and your plot will show up!

- Plotting from an IPython shell: This is actually one of the most convenient ways to run Matplotlib interactively. To make the plots show up, you need to use what is called the %matplotlib magic command after starting IPython:

    ```
    In [1]: %matplotlib
    Using matplotlib backend: TkAgg
    In [2]: import matplotlib.pyplot as plt
    ```

Then, all your plots will show up automatically without having to call plt.show() every time.

- Plotting from a Jupyter Notebook: If you are viewing this code from a browser-based Jupyter Notebook, you need to use the same %matplotlib magic. However, you also have the option of embedding graphics directly in the notebook, with two possible outcomes:
    - %matplotlib notebook will lead to interactive plots embedded within the notebook
    - %matplotlib inline will lead to static images of your plots embedded within the notebook

In this book, we will generally opt for the inline option:

```
In [6]: %matplotlib inline
```

Now let's try this again:

```
In [7]: plt.plot(x, np.sin(x))
Out[7]: [<matplotlib.lines.Line2D at 0x7f3aac426eb8>]
```

The preceding command gives the following output:

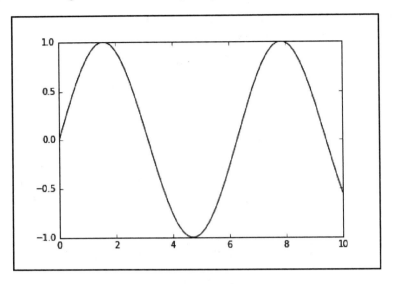

Plotting the sin function using Matplotlib

If you want to save the figure for later, you have the option to do so directly from within IPython or Jupyter Notebook:

```
In [8]: savefig('figures/02.04-sine.png')
```

Just make sure that you use a supported file ending, such as .jpg, .png, .tif, .svg, .eps, or .pdf.

You can change the style of your plots by running plt.style.use(style_name) after importing Matplotlib. All available styles are listed in plt.style.available. For example, try plt.style.use('fivethirtyeight'), plt.style.use('ggplot'), or plt.style.use('seaborn-dark'). For added fun, run plt.xkcd(), then try and plot something else.

Visualizing data from an external dataset

As a final test for this chapter, let's visualize some data from an external dataset, such as the digits dataset from scikit-learn.

Specifically, We will need three tools for visualization:

- scikit-learn for the actual data
- NumPy for data munging
- Matplotlib

So let's start by importing all of these:

```
In [1]: import numpy as np
   ...      from sklearn import datasets
   ...      import matplotlib.pyplot as plt
   ...      %matplotlib inline
```

The first step is to actually load the data:

```
In [2]: digits = datasets.load_digits()
```

If we remember correctly, `digits` is supposed to have two different fields: a `data` field containing the actual image data, and a `target` field containing the image labels. Rather than trust our memory, we should simply investigate the digits object. We do this by typing out its name, adding a period, and then hitting the *TAB* key: `digits.<TAB>`. This will reveal that the `digits` object also contains some other fields, such as one called `images`. The two fields, `images` and `data`, seem to simply differ by shape:

```
In [3]: print(digits.data.shape)
   ...      print(digits.images.shape)
Out[3]: (1797, 64)
   ...      (1797, 8, 8)
```

In both cases, the first dimension corresponds to the number of images in the dataset. However, `data` has all the pixels lined up in one big vector, whereas `images` preserves the 8 x 8 spatial arrangement of each image.

Thus, if we wanted to plot a single image, the `images` field would be more appropriate. First, we grab a single image from the dataset using NumPy's array slicing:

```
In [4]: img = digits.images[0, :, :]
```

Here, we are saying that we want to grab the first row in the 1,797 items-long array and all the corresponding 8 x 8=64 pixels. We can then plot the image using `plt`'s `imshow` function:

```
In [5]: plt.imshow(img, cmap='gray')
Out[5]: <matplotlib.image.AxesImage at 0x7efcd27f30f0>
```

The preceding command gives the following output:

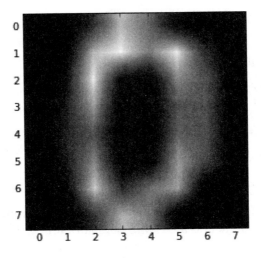

An example image from the digits dataset

In addition, I also specified a color map with the `cmap` argument. By default, Matplotlib uses MATLAB's default colormap **jet**. However, in the case of grayscale images, the **gray** colormap makes more sense.

Finally, we can plot a whole number of digit samples using `plt`'s `subplot` function. The `subplot` function is the same as in MATLAB, where we specify the number of rows, number of columns, and current subplot index (starts counting at 1). We will use `for loop` to iterate over the first ten images in the dataset and every image gets assigned its own subplot:

```
In [6]: for image_index in range(10):
   ...      # images are 0-indexed, but subplots are 1-indexed
   ...      subplot_index = image_index + 1
   ...      plt.subplot(2, 5, subplot_index)
   ...      plt.imshow(digits.images[image_index, :, :], cmap='gray')
```

This leads to the following output:

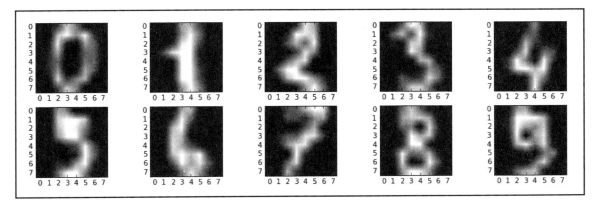

Ten example images from the digits database

 Another great resource for all sorts of datasets is the Machine Learning Repository of my *Alma Mater*, the University of California, Irvine: http://archive.ics.uci.edu/ml.

Dealing with data using OpenCV's TrainData container in C++

For the sake of completeness and for those who insist on using the C++ API of OpenCV, let's do a quick detour on OpenCV's `TrainData` container that allows us to load numerical data from .csv files.

Among other things, in C++ the `ml` module contains a class called `TrainData`, which provides a container to work with data in C++. Its functionality is limited to reading (preferably) numerical data from .csv files (containing comma-separated values). Hence, if the data that you want to work with comes in a neatly organized .csv file, this class will save you a lot of time. If your data comes from a different source, I'm afraid your best option might be to create a .csv file by hand, using a suitable program such as Open Office or Microsoft Excel.

The most important method of the `TrainData` class is called `loadFromCSV`, which accepts the following parameters:

- `const String& filename`: The input filename
- `int headerLineCount`: The number of lines in the beginning to skip
- `int responseStartIdx`: The index of the first output variable
- `int responseEndIdx`: The index of the last output variable
- `const String& varTypeSpec`: A text string describing the data type of all output variables
- `char delimiter`: The character used to separate values in each line
- `char missch`: The character used to specify missing measurements

If you have some nice all-float data that lives in a comma-separated file, you can load it as follows:

```
Ptr<TrainData> tDataContainer = TrainData::loadFromCSV("file.csv",
                         0,    // number of lines to skip
                         0);   // index of 1st and only output var
```

The class provides a number of handy functions to split the data into training and test sets and access individual data points in the training or test set. For example, if your file contains 100 samples, you could assign the first 90 samples to the training set and leave the remaining 10 to the test set. First, it might be a good idea to shuffle the samples:

```
tDataContainer->shuffleTrainTest();
tDataContainer->setTrainTestSplit(90);
```

Then it is easy to store all training samples in an OpenCV matrix:

```
cv::Mat trainData = tDataContainer->getTrainSamples();
```

You can find all relevant functions described here:
http://docs.opencv.org/3.1.0/dc/d32/classcv_1_1ml_1_1TrainData.html.

Other than that, I'm afraid the `TrainData` container and its use cases might be a bit antiquated. So for the rest of the book, we will focus on Python instead.

Summary

In this chapter, we talked about a typical workflow to deal with machine learning problems: how we can extract informative features from raw data, how we can use data and labels to train a machine learning model, and how we can use the finalized model to predict new data labels. We learned that it is essential to split data into a training set and test set, as this is the only way to know how well a model will generalize to new data points.

On the software side of things, we significantly improved our Python skills. We learned how to use NumPy arrays to store and manipulate data and how to use Matplotlib for data visualization. We talked about scikit-learn and its many useful data resources. Finally, we also addressed OpenCV's own `TrainData` container, which provides some relief for users of OpenCV's C++ API.

With these tools in hand, we are now ready to implement our first real machine learning model! In the next chapter, we will focus on supervised learning and its two main problem categories, classification and regression.

3
First Steps in Supervised Learning

This is the moment you've been waiting for, isn't it?

We have covered all the bases-we have a functioning Python environment, we have OpenCV installed, and we know how to handle data in Python. Now it's time to build our first machine learning system! And what better way to start off than to focus on one of the most common and successful types of machine learning: **supervised learning**?

From the previous chapter, we already know that supervised learning is all about learning regularities in some training data by using the labels that come with it so that we can predict the labels of some new, never-seen-before test data. In this chapter, we want to dig a little deeper, and learn how to turn our theoretical knowledge into something practical.

Specifically, we want to address the following questions:

- What's the difference between **classification** and **regression**, and when do I use which?
- What is a **k-nearest neighbor** (*k*-**NN**) classifier, and how do I implement one in OpenCV?
- How do I use **logistic regression** for classification, and why is it named so confusingly?
- How do I build a **linear regression** model in OpenCV, and how does it differ from **Lasso** and **ridge regression**?

Let's jump right in!

Understanding supervised learning

We have previously established that the goal of supervised learning is always to predict the **labels** (or **target values**) of some data. However, depending on the nature of these labels, supervised learning can come in two distinct forms:

- **Classification**: Supervised learning is called classification whenever we use the data to predict categories. A good example of this is when we try to predict whether an image contains a **cat** or a **dog**. Here, the labels of the data are categorical, either one or the other, but never a mixture of categories. For example, a picture contains either a cat or a dog, never 50 percent cat and 50 percent dog (before you ask, no, here we do not consider pictures of the cartoon character CatDog), and our job is simply to tell which one it is. When there are only two choices, it is called **two-class** or **binary** classification. When there are more than two categories, as when predicting what the weather will be like the next day, it is known as **multi-class** classification.
- **Regression**: Supervised learning is called regression whenever we use the data to predict real values. A good example of this is when we try to predict stock prices. Rather than predicting stock categories, the goal of regression is to predict a **target value** as accurately as possible; for example, to predict the stock prices with as little an error as possible.

Perhaps the easiest way to figure out whether we are dealing with a classification or regression problem is to ask ourselves the following question: What are we actually trying to predict? The answer is given in the following figure:

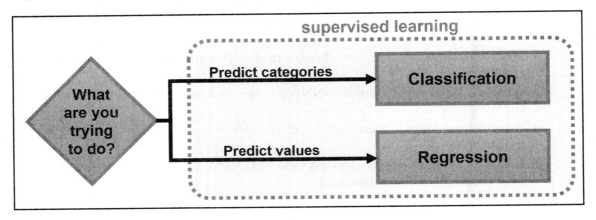

Differentiating between classification and regression problems

Chapter 3

Having a look at supervised learning in OpenCV

Knowing how supervised learning works is pretty if we can't put it into practice. Thankfully, OpenCV provides a pretty straightforward interface for all its **statistical learning models**, which includes all supervised learning models.

In OpenCV, every machine learning model derives from the `cv::ml::StatModel` **base class**. This is fancy talk for saying that if we want to be a machine learning model in OpenCV, we have to provide all the functionality that `StatModel` tells us to. This includes a **method** to train the model (called `train`) and a method to measure the performance of the model (called `calcError`).

> In **object-oriented programming (OOP)**, functions are often called **objects** or **classes**. An object can itself consist of a number of functions, called **methods**, as well as variables, called **members** or **attributes**. You can learn more about OOP in Python at https://docs.python.org/3/tutorial/classes.html.

Thanks to this organization of the software, setting up a machine learning model in OpenCV always follows the same logic:

- **Initialization**: We call the model by name to create an empty instance of the model.
- **Set parameters**: If the model needs some parameters, we can set them via setter methods, which can be different for every model. For example, in order for a k-NN algorithm to work, we need to specify its open parameter, k (as we will find out later).
- **Train the model**: Every model must provide a method called `train`, used to fit the model to some data.
- **Predict new labels**: Every model must provide a method called `predict`, used to predict the labels of new data.
- **Score the model**: Every model must provide a method called `calcError`, used to measure performance. This calculation might be different for every model.

> Because OpenCV is a vast and community-driven project, not every algorithm follows these rules to the extent that we as users might expect. For example, the k-NN algorithm does most of its work in a `findNearest` method, although `predict` still works. We will make sure to point out these discrepancies as we work through different examples.

As we will make the occasional use of scikit-learn to implement some machine learning algorithms that OpenCV does not provide, it is worth pointing out that learning algorithms in scikit-learn follow an almost identical logic. The most notable difference is that scikit-learn sets all the required model parameters in the initialization step. In addition, it calls the training function `fit` instead of `train`, and the scoring function `score` instead of `calcError`.

Measuring model performance with scoring functions

One of the most important parts of building a machine learning system is to find a way to measure the quality of the model predictions. In real-life scenarios, a model will rarely get everything right. From earlier chapters, we know that we are supposed to use data from the test set to evaluate our model. But how exactly does that work?

The short, but not very helpful, answer is that it depends on the model. People have come up with all sorts of **scoring functions** that can be used to evaluate our model in all possible scenarios. The good news is that a lot of them are actually part of scikit-learn's `metrics` module.

Let's have a quick look at some of the most important scoring functions.

Scoring classifiers using accuracy, precision, and recall

In a binary classification task, where there are only two different class labels, there are a number of different ways to measure classification performance. Some common metrics are as follows:

- `accuracy_score`: **Accuracy** counts the number of data points in the test set that have been predicted correctly, and returns that number as a fraction of the test set size. Sticking to the example of classifying pictures as cats or dogs, accuracy indicates the fraction of pictures that have been correctly classified as containing either a cat or a dog. This is the most basic scoring function for classifiers.
- `precision_score`: **Precision** describes the ability of a classifier not to label as cat a picture that contains a dog. In other words, out of all the pictures in the test set that the classifier thinks contain a cat, precision is the fraction of pictures that actually contain a cat.

- `recall_score`: **Recall** (or **sensitivity**) describes the ability of a classifier to retrieve all the pictures that contain a cat. In other words, out of all the pictures of cats in the test set, recall is the fraction of pictures that have been correctly identified as pictures of cats.

Let's say, we have some `ground truth` class labels that are either zeros or ones. We can generate them at random using NumPy's **random number generator**. Obviously, this means that whenever we rerun our code, new data points will be generated at random. However, for the purpose of this book, this is not very helpful, as I want you to be able to run the code and always get the same result as me. A nice trick to get that is to fix the **seed** of the random number generator. This will make sure the generator is initialized the same way every time you run the script.

We can fix the seed of the random number generator using the following code:

```
In [1]: import numpy as np
In [2]: np.random.seed(42)
```

Then we can generate five random labels that are either zeros or ones by picking random integers in the range (0, 2):

```
In [3]: y_true = np.random.randint(0, 2, size=5)
   ...: y_true
Out[3]: array([0, 1, 0, 0, 0])
```

In the literature, these two classes are sometimes also called **positives** (all data points with class label 1) and **negatives** (all other data points).

Let's assume we have a classifier that tries to predict the class labels mentioned earlier. For the sake of argument, let's say the classifier is not very smart, and always predicts label 1. We can mock this behavior by hard-coding the prediction labels:

```
In [4]: y_pred = np.ones(5, dtype=np.int32)
   ...: y_pred
Out[4]: array([1, 1, 1, 1, 1])
```

What is the accuracy of our prediction?

As mentioned earlier, accuracy counts the number of data points in the test set that have been predicted correctly, and returns that number as a fraction of the test set size. We correctly predicted only the second data point (where the true label is 1). In all other cases, the true label was a 0, yet we predicted 1. Hence, our accuracy should be 1/5 or 0.2.

A naive implementation of an accuracy metric might sum up all occurrences where the predicted class label matched the true class label:

```
In [5]: np.sum(y_true == y_pred) / len(y_true)
Out[5]: 0.20000000000000001
```

Close enough, Python.

A smarter, and more convenient, implementation is provided by scikit-learn's `metrics` module:

```
In [6]: from sklearn import metrics
In [7]: metrics.accuracy_score(y_true, y_pred)
Out[7]: 0.20000000000000001
```

That wasn't too hard, was it? However, in order to understand precision and recall, we need a general understanding of **type I** and **type II errors**. Let's recall that data points with class label 1 are often called positives, and data points with class label 0 (or -1) are often called negatives. Then classifying a specific data point can have one of four possible outcomes, as illustrated with the following confusion matrix:

	is truly positive	is truly negative
predicted positive	true positive	false positive
predicted negative	false negative	true negative

Let's break this down. If a data point was truly a positive, and we predicted a positive, we got it all right! In this case, the outcome is called a **true positive**. If we thought the data point was a positive, but it was really a negative, we falsely predicted a positive (hence the term, **false positive**). Analogously, if we thought the data point was a negative, but it was really a positive, we falsely predicted a negative (**false negative**). Finally, if we predicted a negative and the data point was truly a negative, we found a **true negative**.

In statistical hypothesis testing, false positives are also known as **type I errors** and false negatives are also known as **type II errors**.

Let's quickly calculate these four metrics on our mock-up data. We have a true positive, where the true label is a 1 and we also predicted a 1:

```
In [8]:   truly_a_positive = (y_true == 1)
In [9]:   predicted_a_positive = (y_pred == 1)
In [10]:  true_positive = np.sum(predicted_a_positive * truly_a_positive )
...       true_positive
Out[10]: 1
```

Similarly, a false positive is where we predicted a 1 but the `ground truth` was really a 0:

```
In [11]:  false_positive = np.sum((y_pred == 1) * (y_true == 0))
...       false_positive
Out[11]: 4
```

I'm sure by now you've got the hang of it. But do we even have to do math in order to know about predicted negatives? Our not-so-smart classifier never predicted 0, so `(y_pred == 0)` should never be true:

```
In [12]:  false_negative = np.sum((y_pred == 0) * (y_true == 1))
...       false_negative
Out[12]: 0
In [13]:  true_negative = np.sum((y_pred == 0) * (y_true == 0))
...       true_negative
Out[13]: 0
```

To make sure we did everything right, let's calculate accuracy one more time. Accuracy should be the number of true positives plus the number of true negatives (that is, everything we got right) divided by the total number of data points:

```
In [14]:  accuracy = (true_positive + true_negative) / len(y_true)
...       accuracy
Out[14]: 0.20000000000000001
```

Success! Precision is then given as the number of true positives divided by the number of all true predictions:

```
In [15]: precision = true_positive / (true_positive + true_negative)
    ...  precision
Out[15]: 1.0
```

Turns out that precision isn't better than accuracy in our case. Let's check our math with scikit-learn:

```
In [16]: metrics.precision_score(y_true, y_pred)
Out[16]: 0.20000000000000001
```

Finally, `recall` is given as the fraction of all positives that we correctly classified as positives:

```
In [17]: recall = true_positive / (true_positive + false_negative)
    ...  recall
Out[17]: 1.0
In [18]: metrics.recall_score(y_true, y_pred)
Out[18]: 1.0
```

Perfect recall! But, going back to our mock-up data, it should be clear that this excellent recall score was mere luck. Since there was only a single 1 in our mock-up dataset, and we happened to correctly classify it, we got a perfect recall score. Does that mean our classifier is perfect? Not really! But we have found three useful metrics that seem to measure complementary aspects of our classification performance.

Scoring regressors using mean squared error, explained variance, and R squared

When it comes to regression models, our metrics, as shown earlier, don't work anymore. After all, we are now predicting continuous output values, not distinct classification labels. Fortunately, scikit-learn provides some other useful scoring functions:

- `mean_squared_error`: The most commonly used error metric for regression problems is to measure the **squared error** between the predicted and the true target value for every data point in the training set, averaged across all the data points.

- `explained_variance_score`: A more sophisticated metric is to measure to what degree a model can explain the variation or dispersion of the test data. Often, the amount of **explained variance** is measured using the **correlation coefficient**.
- `r2_score`: The R^2 **score** (pronounced **R squared**) is closely related to the explained variance score, but uses an unbiased variance estimation. It is also known as the **coefficient of determination**.

Let's create another mock-up dataset. Let's say we are observing data that looks like a sin as a function of x values. We start by generating 100 equally spaced x values between 0 and 10:

```
In [19]: x = np.linspace(0, 10, 100)
```

However, real data is always noisy. To honor this fact, we want the target values `y_true` to be noisy, too. We do this by adding noise to the `sin` function:

```
In [20]: y_true = np.sin(x) + np.random.rand(x.size) - 0.5
```

Here, we use NumPy's `rand` function to add noise in the range [0, 1], but then center the noise around 0 by subtracting 0.5. Hence, we effectively jitter every data point either up or down by a maximum of 0.5.

Let's assume our model was smart enough to figure out the `sin(x)` relationship. Hence, the predicted *y* values are given as follows:

```
In [21]: y_pred = np.sin(x)
```

What does this data look like? We can use **Matplotlib** to visualize them:

```
In [22]: import matplotlib.pyplot as plt
    ...: plt.style.use('ggplot')
    ...: %matplotlib inline
In [23]: plt.plot(x, y_pred, linewidth=4, label='model')
    ...: plt.plot(x, y_true, 'o', label='data')
    ...: plt.xlabel('x')
    ...: plt.ylabel('y')
    ...: plt.legend(loc='lower left')
Out[23]: <matplotlib.legend.Legend at 0x265fbeb9f98>
```

This will produce the following line plot:

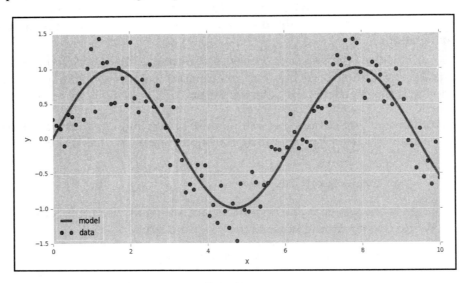

Predicted y-values and ground truth data

The most straightforward metric to determine how good our model predictions are is the mean squared error. For each data point, we look at the difference between the predicted and the actual *y* value, and then square it. We then compute the average of this squared error over all the data points:

```
In [24]: mse = np.mean((y_true - y_pred) ** 2)
   ...       mse
Out[24]: 0.085318394808423778
```

For our convenience, scikit-learn provides its own implementation of the mean squared error:

```
In [25]: metrics.mean_squared_error(y_true, y_pred)
Out[25]: 0.085318394808423778
```

Another common metric is to measure the scatter or variation in the data: if every data point was equal to the mean of all the data points, we would have no scatter or variation in the data, and we could predict all future data points with a single data value. This would be the world's most boring machine learning problem. Instead, we find that the data points often follow some unknown, hidden relationship, that we would like to uncover. In the previous example, this would be the *y=sin(x)* relationship, which causes the data to be scattered.

We can measure how much of that scatter in the data (or **variance**) we can explain. We do this by calculating the variance that still exists between the predicted and the actual labels; this is all the variance our predictions could not explain. If we normalize this value by the total variance in the data, we get what is known as the **fraction of variance unexplained**:

```
In [26]: fvu = np.var(y_true - y_pred) / np.var(y_true)
   ...   fvu
Out[26]: 0.16397032626629501
```

Because this metric is a fraction, its values must lie between 0 and 1. We can subtract this fraction from 1 to get the fraction of variance explained:

```
In [27]: fve = 1.0 - fvu
   ...   fve
Out[27]: 0.83602967373370496
```

Let's verify our math with scikit-learn:

```
In [28]: metrics.explained_variance_score(y_true, y_pred)
Out[28]: 0.83602967373370496
```

Spot on! Finally, we can calculate what is known as the coefficient of determination, or R^2 (pronounced R squared). R^2 is closely related to the fraction of variance explained, and compares the mean squared error calculated earlier to the actual variance in the data:

```
In [29]: r2 = 1.0 - mse / np.var(y_true)
   ...   r2
Out[29]: 0.8358169419264746
```

The same value can be obtained with scikit-learn:

```
In [30]: metrics.r2_score(y_true, y_pred)
Out[30]: 0.8358169419264746
```

The better our predictions fit the data, in comparison to taking the simple average, the closer the value of the R^2 score will be to 1. The R^2 score can take on negative values, as model predictions can be arbitrarily worse than 1. A constant model that always predicts the expected value of y, independent of the input x, would get a R^2 score of 0:

```
In [31]: metrics.r2_score(y_true, np.mean(y_true) * np.ones_like(y_true))
Out[31]: 0.0
```

Using classification models to predict class labels

With these tools in hand, we can now take on our first real classification example.

Consider the small town of **Randomville**, where people are crazy about their two sports teams, the **Randomville Reds** and the **Randomville Blues**. The Reds had been around for a long time, and people loved them. But then some out-of-town millionaire came along and bought the Reds' top scorer and started a new team, the Blues. To the discontent of most Reds fans, the top scorer would go on to win the championship title with the Blues. Years later he would return to the Reds, despite some backlash from fans who could never forgive him for his earlier career choices. But anyway, you can see why fans of the Reds don't necessarily get along with fans of the Blues. In fact, these two fan bases are so divided that they never even live next to each other. I've even heard stories where the Red fans deliberately moved away once Blues fans moved in next door. True story!

Anyway, we are new in town and are trying to sell some Blues merchandise to people by going from door to door. However, every now and then we come across a bleeding-heart Reds fan who yells at us for selling Blues stuff and chases us off their lawn. Not nice! It would be much less stressful, and a better use of our time, to avoid these houses altogether and just visit the Blues fans instead.

Confident that we can learn to predict where the Reds fans live, we start keeping track of our encounters. If we come by a Reds fan's house, we draw a red triangle on we handy town map; otherwise, we draw a blue square. After a while, we get a pretty good idea of where everyone lives:

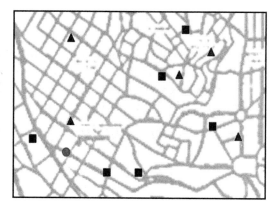

Randomville's town map

However, now we approach the house that is marked as a green circle in the preceding map. Should we knock on their door? We try to find some clue as to what team they prefer (perhaps a team flag hanging from the back porch), but we can't see any. How can we know if it is safe to knock on their door?

What this silly example illustrates is exactly the kind of problem a supervised learning algorithm can solve. We have a bunch of observations (houses, their locations, and their color) that make up our training data. We can use this data to learn from experience, so that when we face the task of predicting the color of a new house, we can make a well-informed estimate.

As we mentioned earlier, fans of the Reds are really passionate about their team, so they would never move next to a Blues fan. Couldn't we use this information and look at all the neighboring houses, in order to find out what kind of fan lives in the new house?

This is exactly what the *k*-NN algorithm would do.

Understanding the k-NN algorithm

The *k*-NN algorithm is arguably one of the simplest machine learning algorithms. The reason for this is that we basically only need to store the training dataset. Then, in order to make a prediction for a new data point, we only need to find the closest data point in the training dataset-its **nearest neighbor**.

In a nutshell, the *k*-NN algorithm argues that a data point probably belongs to the same class as its neighbors. Think about it: if our neighbor is a Reds fan, we're probably Reds fans, too; otherwise we would have moved away a long time ago. The same can be said for the Blues.

Of course, some neighborhoods might be a little more complicated. In this case, we would not just consider our closest neighbor (where *k=1*), but instead our *k* nearest neighbors. To stick with our example as mentioned earlier, if we were Reds fans, we probably wouldn't move into a neighborhood where the majority of people are Blues fans.

That's all there is to it.

Implementing k-NN in OpenCV

Using OpenCV, we can easily create a *k*-NN model via the `cv2.ml.KNearest_create()` function. Building the model then involves the following steps:

1. Generate some training data.
2. Create a *k*-NN object for a given number *k*.
3. Find the *k* nearest neighbors of a new data point that we want to classify.
4. Assign the class label of the new data point by majority vote.
5. Plot the result.

We first import all the necessary modules: OpenCV for the *k*-NN algorithm, NumPy for data munging, and Matplotlib for plotting. If you are working in a Jupyter Notebook, don't forget to call the `%matplotlib inline` magic:

```
In [1]: import numpy as np
   ...  import cv2
   ...  import matplotlib.pyplot as plt
   ...  %matplotlib inline
In [2]: plt.style.use('ggplot')
```

Generating the training data

The first step is to generate some training data. For this, we will use NumPy's random number generator. As discussed in the previous section, we will fix the seed of the random number generator, so that re-running the script will always generate the same values:

```
In [3]: np.random.seed(42)
```

Alright, now let's get to it. What should our training data look like exactly?

In the previous example, each data point is a house on the town map. Every data point has two features (that is, the *x* and *y* coordinates of its location on the town map) and a class label (that is, a blue square if a Blues fan lives there, and a red triangle if a Reds fan lives there). The features of a single data point can therefore be represented by a two-element vector holding its *x* and *y* coordinates on the town map. Similarly, its label is the number 0 if it is a blue square, or the number 1 if it is a red triangle.

We can generate a single data point by picking random locations on the map and a random label (either 0 or 1). Let's say the town map spans a range $0 <= x < 100$ and $0 <= y < 100$. Then we can generate a random data point as follows:

```
In [4]: single_data_point = np.random.randint(0, 100, 2)
   ...  single_data_point
Out[4]: array([51, 92])
```

As shown in the preceding output, this will pick two random integers between 0 and 100. We will interpret the first integer as the data point's *x* coordinate on the map, and the second integer as the point's *y* coordinate. Similarly, let's pick a label for the data point:

```
In [5]: single_label = np.random.randint(0, 2)
   ...  single_label
Out[5]: 0
```

Turns out that this data point would have class 0, which we interpret as a blue square.

Let's wrap this process in a function that takes as input the number of data points to generate (that is, num_samples) and the number of features every data point has (that is, num_features):

```
In [6]: def generate_data(num_samples, num_features=2):
   ...      """Randomly generates a number of data points"""
```

Since in our case the number of features is two, it is okay to use this number as a default argument value. This way, if we don't explicitly specify num_features when calling the function, a value of 2 is automatically assigned to it. I'm sure you already knew that.

The data matrix we want to create should have num_samples rows and num_features columns, and every element in the matrix should be an integer drawn randomly from the range [0, 100]:

```
   ...          data_size = (num_samples, num_features)
   ...          data = np.random.randint(0, 100, size=data_size)
```

Similarly, we want to create a vector that contains a random integer label in the range [0, 2] for all samples:

```
   ...          labels_size = (num_samples, 1)
   ...          labels = np.random.randint(0, 2, size=labels_size)
```

First Steps in Supervised Learning

Don't forget to have the function return the generated data:

```
...         return data.astype(np.float32), labels
```

 OpenCV can be a bit finicky when it comes to data types, so make sure to always convert your data points to `np.float32`!

Let's put the function to test and generate an arbitrary number of data points, let's say eleven, whose coordinates are chosen randomly:

```
In [7]: train_data, labels = generate_data(11)
   ...: train_data
Out[7]: array([[ 71.,  60.],
               [ 20.,  82.],
               [ 86.,  74.],
               [ 74.,  87.],
               [ 99.,  23.],
               [  2.,  21.],
               [ 52.,   1.],
               [ 87.,  29.],
               [ 37.,   1.],
               [ 63.,  59.],
               [ 20.,  32.]], dtype=float32)
```

As we can see from the preceding output, the `train_data` variable is an 11 x 2 array, where each row corresponds to a single data point. We can also inspect the first data point with its corresponding label by indexing into the array:

```
In [8]: train_data[0], labels[0]
Out[8]: (array([ 71.,  60.], dtype=float32), array([0]))
```

This tells us that the first data point is a blue square (because it has class 0) and lives at location (x, y) = (71, 60) on the town map. If we want, we can plot this data point on the town map using Matplotlib:

```
In [9]: plt.plot(train_data[0, 0], train_data[0, 1], 'sb')
   ...: plt.xlabel('x coordinate')
   ...: plt.ylabel('y coordinate')
Out[9]: [<matplotlib.lines.Line2D at 0x137814226a0>]
```

[62]

But what if we want to visualize the whole training set at once? Let's write a function for that. The function should take as input a list of all the data points that are blue squares (`all_blue`) and a list of the data points that are red triangles (`all_red`):

```
In [10]: def plot_data(all_blue, all_red):
```

Our function should then plot all the blue data points as blue squares (using color `'b'` and marker `'s'`), which we can achieve with the `scatter` function of Matplotlib. For this to work, we have to pass the blue data points as an *N x 2* array, where *N* is the number of samples. Then `all_blue[:, 0]` contains all the *x* coordinates of the data points, and `all_blue[:, 1]` contains all the *y* coordinates:

```
    ...       plt.scatter(all_blue[:, 0], all_blue[:, 1], c='b',
                          marker='s', s=180)
```

Analogously, the same can be done for all the red data points:

```
    ...       plt.scatter(all_red[:, 0], all_red[:, 1], c='r',
                          marker='^', s=180)
```

Finally, we annotate the plot with labels:

```
    ...       plt.xlabel('x coordinate (feature 1)')
    ...       plt.ylabel('y coordinate (feature 2)')
```

Let's try it on our dataset! First we have to split all the data points into red and blue sets. We can quickly select all the elements of the `labels` array created earlier that are equal to 0, using the following command (where `ravel` flattens the array):

```
In [11]: labels.ravel() == 0
Out[11]: array([ True,  True, False, False, False,  True, False,  True,
                 True,  True], dtype=bool)
```

All the blue data points are then all the rows of the `train_data` array created earlier, whose corresponding label is 0:

```
In [12]: blue = train_data[labels.ravel() == 0]
```

The same can be done for all the red data points:

```
In [13]: red = train_data[labels.ravel() == 1]
```

Finally, let's plot all the data points:

```
In [14]: plot_data(blue, red)
```

This will create the following figure:

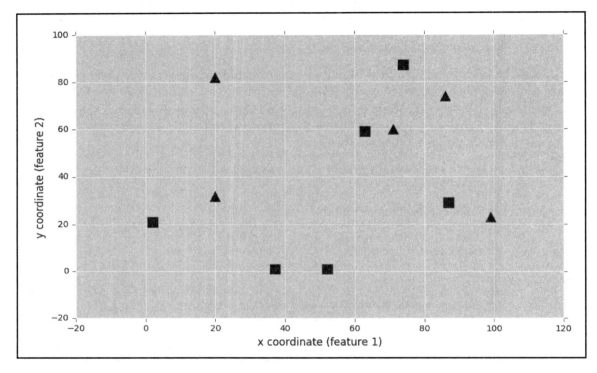

Visualizing the whole training dataset

Training the classifier

Now it's time to train the classifier.

As with all other machine learning functions, the k-NN classifier is part of OpenCV 3.1's `ml` module. We can create a new classifier using the following command:

```
In [15]: knn = cv2.ml.KNearest_create()
```

 In the older versions of OpenCV, this function might be called `cv2.KNearest()` instead.

We then pass our training data to the `train` method:

```
In [16]: knn.train(train_data, cv2.ml.ROW_SAMPLE, labels)
Out[16]: True
```

Here, we have to tell `knn` that our data is an *N x 2* array (that is, every row is a data point). Upon success, the function returns `True`.

Predicting the label of a new data point

The other really helpful method that `knn` provides is called `findNearest`. It can be used to predict the label of a new data point based on its nearest neighbors.

Thanks to our `generate_data` function, it is actually really easy to generate a new data point! We can think of a new data point as a dataset of size 1:

```
In [17]: newcomer, _ = generate_data(1)
Out[17]: newcomer
```

Our function also returns a random label, but we are not interested in that. Instead, we want to predict it using our trained classifier! We can tell Python to ignore an output value with an underscore (_).

Let's have a look at our town map again. We will plot the training set as we did earlier, but also add the new data point as a green circle (since we don't know yet whether it is supposed to be a blue square or a red triangle):

```
In [18]: plot_data(blue, red)
    ...:     plt.plot(newcomer[0, 0], newcomer[0, 1], 'go', markersize=14);
```

 You can add a semicolon to the `plt.plot` function call in order to suppress its output, the same as in Matlab.

[65]

The preceding code will produce the following figure (minus the rings):

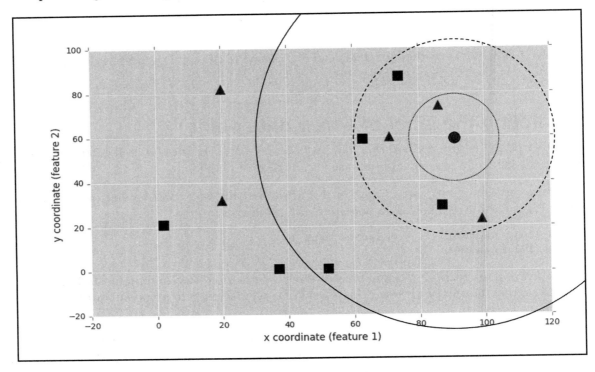

The entire training set, plus a new data point (green) whose label has yet to be determined

If you had to guess based on its neighbors, what label would you assign the new data point- blue or red?

Well, it depends, doesn't it? If we look at the house closest to it (the one living roughly at (*x*, *y*) = (85, 75), circled with a dotted line in the preceding figure), we would probably assign the new data point to be a red triangle as well. This is exactly what our classifier would predict for *k=1*:

```
In [19]: ret, results, neighbor, dist = knn.findNearest(newcomer, 1)
   ...:  print("Predicted label:\t", results)
   ...:  print("Neighbor's label:\t", neighbor)
   ...:  print("Distance to neighbor:\t", dist)
Out[19]: Predicted label:         [[ 1.]]
         Neighbor's label:        [[ 1.]]
         Distance to neighbor:    [[ 250.]]
```

Here, `knn` reports that the nearest neighbor is 250 arbitrary units away, that the neighbor has label 1 (which we said corresponds to red triangles), and that therefore the new data point should also have label 1. The same would be true if we looked at the $k=2$ nearest neighbors, and the $k=3$ nearest neighbors. But we want to be careful not to pick arbitrary even numbers for k. Why is that? Well, you can see it in the preceding figure (dashed circle); among the six nearest neighbors within the dashed circle, there are three blue squares and three red triangles—we have a tie!

In the case of a tie, OpenCV's implementation of k-NN will prefer the neighbors with a closer overall distance to the data point.

Finally, what would happen if we dramatically widened our search window and classified the new data point based on its $k=7$ nearest neighbors (circled with a solid line in the figure mentioned earlier)?

Let's find out by calling the `findNearest` method with $k=7$ neighbors:

```
In [20]: ret, results, neighbors, dist = knn.findNearest(newcomer, 7)
    ...: print("Predicted label:\t", results)
    ...: print("Neighbors' labels:\t", neighbors)
    ...: print("Distance to neighbors:\t", dist)
Out[20]: Predicted label:        [[ 0.]]
         Neighbors' label:       [[ 1.  1.  0.  0.  0.  1.  0.]]
         Distance to neighbors:  [[ 250.  401.  784.  916.  1073.  1360.  4885.]]
```

Suddenly, the predicted label is 0 (blue square). The reason is that we now have four neighbors within the solid circle that are blue squares (label 0), and only three that are red triangles (label 1). So the majority vote would suggest making the newcomer a blue square as well.

As you can see, the outcome of k-NN changes with the number k. However, often we do not know beforehand which number k is the most suitable. A naive solution to this problem is just to try a bunch of values for k, and see which one performs best. We will learn more sophisticated solutions in later chapters of this book.

Using regression models to predict continuous outcomes

Now let's turn our attention to a regression problem. As I'm sure you can recite in your sleep by now, regression is all about predicting continuous outcomes rather than predicting discrete class labels.

Understanding linear regression

The easiest regression model is called **linear regression**. The idea behind linear regression is to describe a target variable (such as Boston house pricing) with a **linear combination** of features.

To keep things simple, let's just focus on two features. Let's say we want to predict tomorrow's stock prices using two features: today's stock price and yesterday's stock price. We will denote today's stock price as the first feature f_1, and yesterday's stock price as f_2. Then the goal of linear regression would be to learn two **weight coefficients**, w_1 and w_2, so that we can predict tomorrow's stock price as follows:

$$\hat{y} = w_1 f_1 + w_2 f_2$$

Here, $\hat{y}$ is the **prediction** of tomorrow's ground truth stock price y.

> The special case of having only one feature variable is called **simple linear regression**.

We could easily extend this to feature more stock price samples from the past. If we had M feature values instead of two, we would extend the preceding equation to a sum of M products, so that every feature gets accompanied by a weight coefficient. We can write the resulting equation as follows:

$$\hat{y} = w_1 f_1 + w_2 f_2 + \cdots + w_M f_M = \sum_{j=1}^{M} w_j f_j$$

Let's think about this equation geometrically for a second. In the case of a single feature, f_1, the equation for $\hat{y}$ would become $\hat{y} = w_1 f_1$, which is essentially a **straight line**. In the case of two features, $\hat{y} = w_1 f_1 + w_2 f_2$ would describe a **plane** in the feature space, as illustrated in the following figure:

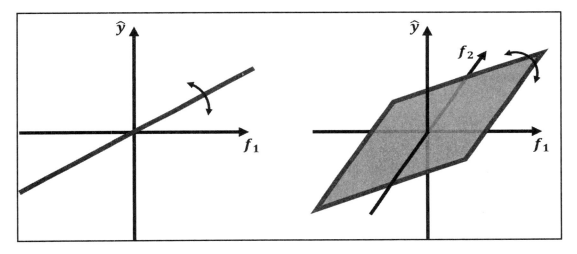

Prediciting target values in two and three dimensions using linear regression

In N dimensions, would become what is known as a **hyperplane**. If a space is N-dimensional, then its hyperplanes have N-1 dimensions.

As is evident in the preceding figure, all of these lines and planes intersect at the **origin**. But, what if the true y values we are trying to approximate don't go through the origin?

In order to be able to offset $\hat{y}$ from the origin, it is customary to add an additional weight coefficient that does not depend on any feature values, and thus acts like a **bias term**. In a 1D case, this term acts as the $\hat{y}$-intercept. In practice, this is often achieved by setting $f_0=1$ so that w_0 can act as the bias term:

$$\hat{y} = w_0 + \sum_{j=1}^{M} w_j f_j$$

Finally, the goal of linear regression is to learn a set of weight coefficients that lead to a prediction that approximates the `ground truth` values as accurately as possible. Rather than explicitly capturing a model's accuracy like we did with classifiers, scoring functions in regression often take the form of so called cost functions (or loss functions).

As discussed earlier in this chapter, there are a number of scoring functions we can use to measure the performance of a regression model. The most commonly used cost function is probably the **mean squared error**, which calculates an error $(y_i - \hat{y}_i)^2$ for every data point i by comparing the prediction $\hat{y}_i$ to the target output value y_i and then taking the average.

Then regression becomes an **optimization problem**—and the task is to find the set of weights that minimizes the cost function.

This is usually done with an **iterative algorithm** that is applied to one data point after the other, thus reducing the cost function step by step. We will talk more deeply about such algorithms in `Chapter 9`, *Using Deep Learning to Classify Handwritten Digits*.

But enough with all this theory—let's do some coding!

Using linear regression to predict Boston housing prices

To get a better understanding of linear regression, we want to build a simple model that can be applied to one of the most famous machine learning datasets known as the **Boston housing prices dataset**. Here, the goal is to predict the value of homes in several Boston neighborhoods in the 1970s, using information such as crime rate, property tax rate, distance to employment centers, and highway accessibility.

Loading the dataset

We can again thank scikit-learn for easy access to the dataset. We first import all the necessary modules, as we did earlier:

```
In [1]: import numpy as np
   ...  from sklearn import datasets
   ...  from sklearn import metrics
   ...  from sklearn import model_selection as modsel
   ...  from sklearn import linear_model
   ...  %matplotlib inline
   ...  import matplotlib.pyplot as plt
   ...  plt.style.use('ggplot')
```

Then, loading the dataset is a one-liner:

```
In [2]: boston = datasets.load_boston()
```

The structure of the `boston` object is identical to the `iris` object, as discussed in the preceding command. We can get more information about the dataset in `'DESCR'`, find all data in `'data'`, all feature names in `'feature_names'`, and all target values in `'target'`:

```
In [3]: dir(boston)
Out[3]: ['DESCR', 'data', 'feature_names', 'target']
```

The dataset contains a total of 506 data points, each of which has 13 features:

```
In [4]: boston.data.shape
Out[4]: (506, 13)
```

Of course, we have only a single target value, which is the housing price:

```
In [5]: boston.target.shape
Out[5]: (506,)
```

Training the model

Believe it or not, OpenCV does not offer any good implementation of linear regression. Some people online say that you can use `cv2.fitLine`, but that is different. This is a perfect opportunity to get familiar with scikit-learn's API:

```
In [6]: linreg = linear_model.LinearRegression()
```

In the preceding command, we want to split the data into training and test sets. We are free to make the split as we see fit, but usually it is a good idea to reserve between 10 percent and 30 percent for testing. Here, we choose 10 percent, using the `test_size` argument:

```
In [7]: X_train, X_test, y_train, y_test = modsel.train_test_split(
   ...         boston.data, boston.target, test_size=0.1,
   ...         random_state=42
   ...     )
```

In scikit-learn, the `train` function is called `fit`, but otherwise behaves exactly the same as in OpenCV:

```
In [8]: linreg.fit(X_train, y_train)
Out[8]: LinearRegression(copy_X=True, fit_intercept=True, n_jobs=1,
                normalize=False)
```

We can look at the mean squared error of our predictions by comparing the true housing prices, `y_train`, to our predictions, `linreg.predict(X_train)`:

```
In [9]: metrics.mean_squared_error(y_train, linreg.predict(X_train))
Out[9]: 22.739484154236614
```

The `score` method of the `linreg` object returns the coefficient of determination (R squared):

```
In [10]: linreg.score(X_train, y_train)
Out[10]: 0.73749340919011974
```

Testing the model

In order to test the **generalization performance** of the model, we calculate the mean squared error on the test data:

```
In [11]: y_pred = linreg.predict(X_test)
In [12]: metrics.mean_squared_error(y_test, y_pred)
Out[12]: 15.010997321630166
```

We note that the mean squared error is a little lower on the test set than the training set. This is good news, as we care mostly about the test error. However, from these numbers, it is really hard to understand how good the model really is. Perhaps it's better to plot the data:

```
In [13]: plt.figure(figsize=(10, 6))
   ...      plt.plot(y_test, linewidth=3, label='ground truth')
   ...      plt.plot(y_pred, linewidth=3, label='predicted')
   ...      plt.legend(loc='best')
```

```
...            plt.xlabel('test data points')
...            plt.ylabel('target value')
Out[13]: <matplotlib.text.Text at 0x7ff46783c7b8>
```

This produces the following figure:

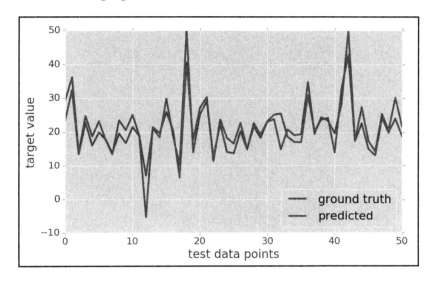

Linear regression model

This makes more sense! Here we see the `ground truth` housing prices for all test samples in blue and our predicted housing prices in red. Pretty close, if you ask me. It is interesting to note though that the model tends to be off the most for really high or really low housing prices, such as the peak values of data point 12, 18, and 42. We can formalize the amount of variance in the data that we were able to explain by calculating R squared:

```
In [14]: plt.plot(y_test, y_pred, 'o')
...            plt.plot([-10, 60], [-10, 60], 'k--')
...            plt.axis([-10, 60, -10, 60])
...            plt.xlabel('ground truth')
...            plt.ylabel('predicted')
```

This will plot the `ground truth` prices, `y_test`, on the x axis, and our predictions, `y_pred`, on the y axis. We also plot a diagonal line for reference (using a black dashed line, `'k--'`), as we will see soon. But we also want to display the R^2 score and mean squared error in a text box:

```
...            scorestr = r'R$^2$ = %.3f' % linreg.score(X_test, y_test)
...            errstr = 'MSE = %.3f' % metrics.mean_squared_error(y_test, y_pred)
...            plt.text(-5, 50, scorestr, fontsize=12)
```

```
...         plt.text(-5, 45, errstr, fontsize=12)
Out[14]: <matplotlib.text.Text at 0x7ff4642d0400>
```

This will produce the following figure, and is a professional way of plotting a model fit:

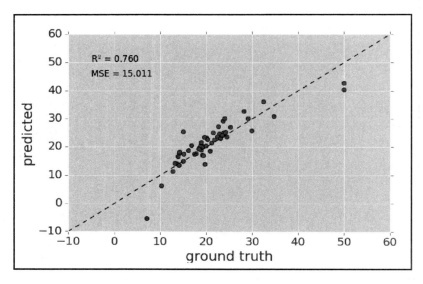

Model predictions versus ground truth

If our model was perfect, then all data points would lie on the dashed diagonal, since `y_pred` would always be equal to `y_true`. Deviations from the diagonal indicate that the model made some errors, or that there is some variance in the data that the model was not able to explain. Indeed, R2 indicates that we were able to explain 76 percent of the scatter in the data, with a mean squared error of **15.011**. These are some hard numbers we can use to compare the linear regression model to some more complicated ones.

Applying Lasso and ridge regression

A common problem in machine learning is that an algorithm might work really well on the training set, but when applied to unseen data it makes a lot of mistakes. You can see how this is problematic, since often we are most interested in how a model **generalizes** to new data. Some algorithms (such as decision trees) are more susceptible to this phenomenon than others, but even linear regression can be affected.

This phenomeon is also known as **overfitting**, and we will talk about it extensively in `Chapter 5`, *Using Decision Trees to Make a Medical Diagnosis*, and `Chapter 11`, *Selecting the Right Model with Hyperparameter Tuning*.

A common technique for reducing overfitting is called **regularization**, which involves adding an additional constraint to the cost function that is independent of all feature values. The two most commonly used **regularizors** are as follows:

- **L1 regularization**: This adds a term to the scoring function that is proportional to the sum of all absolute weight values. In other words, it is based on the **L1 norm** of the weight vector (also known as the **rectilinear distance**, **snake distance**, or **Manhattan distance**). Due to the grid layout of Manhattan's streets, the L1 norm is akin to measuring the distance a New York cab driver covers by driving from point A to B. The resulting algorithm is also known as **Lasso regression**.
- **L2 regularization**: This adds a term to the scoring function that is proportional to the sum of all squared weight values. In other words, it is based on the **L2 norm** of the weight vector (also known as the **Euclidean distance**). Since the L2 norm involves a squaring operation, it punishes strong outliers in the weight vector much harder than the L1 norm. The resulting algorithm is also known as **ridge regression**.

The procedure is exactly the same as the preceding one, but we replace the initialization command to load either a `Lasso` or a `RidgeRegression` object. Specifically, we have to replace the following command:

```
In [6]: linreg = linear_model.LinearRegression()
```

For the Lasso regression algorithm, we would change the preceding line of code to the following:

```
In [6]: lassoreg = linear_model.Lasso()
```

For the ridge regression algorithm, we would change the preceding line of code to the following:

```
In [6]: ridgereg = linear_model.RidgeRegression()
```

I encourage you to test these two algorithms on the Boston dataset in place of conventional linear regression. How does the generalization error (`In [12]`) change? How does the prediction plot (`In [14]`) change? Do you see any improvements in performance?

Classifying iris species using logistic regression

Another famous dataset in the world of machine learning is called the **Iris** dataset. The Iris dataset contains measurements of 150 iris flowers from three different species: **setosa**, **versicolor**, and **viriginica**. These measurements include the length and width of the petals, and the length and width of the sepals, all measured in centimeters:

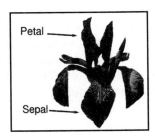

An iris flower

Our goal is to build a machine learning model that can learn the measurements of these iris flowers, whose species are known, so that we can predict the species for a new iris flower.

Understanding logistic regression

Despite its name, **logistic regression** can actually be used as a model for classification. It uses a **logistic function** (or **sigmoid**) to convert any real-valued input x into a predicted output value $\hat{y}$ that takes values between 0 and 1, as shown in the following figure:

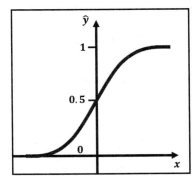

The logistic function

Rounding $\hat{y}$ to the nearest integer effectively classifies the input as belonging either to class 0 or 1.

Of course, most often, our problems have more than one input or feature value, x. For example, the Iris dataset provides a total of four features. For the sake of simplicity, let's focus here on the first two features, sepal length—which we will call feature f_1—and sepal width—which we will call f_2. Using the tricks we learned when talking about linear regression, we know we can express the input x as a **linear combination** of the two features, f_1 and f_2:

$$x = w_1 f_1 + w_2 f_2$$

However, in contrast to linear regression, we are not done yet. From the previous section we know that the sum of products would result in a real-valued, output—but we are interested in a categorical value, zero or one. This is where the logistic function comes in: it acts as a **squashing function**, σ, that compresses the range of possible output values to the range [0, 1]:

$$\hat{y} = \sigma(x)$$

> Because the output is always between 0 and 1, it can be interpreted as a **probability**. If we only have a single input variable x, the output value $\hat{y}$ can be interpreted as the probability of x belonging to class 1.

Now let's apply this knowledge to the Iris dataset!

Loading the training data

The Iris dataset is included with scikit-learn. We first load all the necessary modules, as we did in our earlier examples:

```
In [1]: import numpy as np
   ...      import cv2
   ...      from sklearn import datasets
   ...      from sklearn import model_selection
   ...      from sklearn import metrics
```

First Steps in Supervised Learning

```
...         import matplotlib.pyplot as plt
...         %matplotlib inline
In [2]: plt.style.use('ggplot')
```

Then, loading the dataset is a one-liner:

```
In [3]: iris = datasets.load_iris()
```

This function returns a dictionary we call `iris`, which contains a bunch of different fields:

```
In [4]: dir(iris)
Out[4]: ['DESCR', 'data', 'feature_names', 'target', 'target_names']
```

Here, all the data points are contained in `'data'`. There are 150 data points, each of which have four feature values:

```
In [5]: iris.data.shape
Out[5]: (150, 4)
```

These four features correspond to the sepal and petal dimensions mentioned earlier:

```
In [6]: iris.feature_names
Out[6]: ['sepal length (cm)',
         'sepal width (cm)',
         'petal length (cm)',
         'petal width (cm)']
```

For every data point, we have a class label stored in `target`:

```
In [7]: iris.target.shape
Out[7]: (150,)
```

We can also inspect the class labels, and find that there is a total of three classes:

```
In [8]: np.unique(iris.target)
Out[8]: array([0, 1, 2])
```

Making it a binary classification problem

For the sake of simplicity, we want to focus on a **binary classification problem** for now, where we only have two classes. The easiest way to do this is to discard all data points belonging to a certain class, such as class label 2, by selecting all the rows that **do not** belong to class 2:

```
In [9]: idx = iris.target != 2
   ...: data = iris.data[idx].astype(np.float32)
   ...: target = iris.target[idx].astype(np.float32)
```

Inspecting the data

Before you get started with setting up a model, it is always a good idea to have a look at the data. We did this earlier for the town map example, so let's continue our streak. Using Matplotlib, we create a **scatter plot** where the color of each data point corresponds to the class label:

```
In [10]: plt.scatter(data[:, 0], data[:, 1], c=target,
                     cmap=plt.cm.Paired, s=100)
   ...:  plt.xlabel(iris.feature_names[0])
   ...:  plt.ylabel(iris.feature_names[1])
Out[10]: <matplotlib.text.Text at 0x23bb5e03eb8>
```

To make plotting easier, we limit ourselves to the first two features (iris.feature_names[0] being the sepal length and iris.feature_names[1] being the sepal width). We can see a nice separation of classes in the following figure:

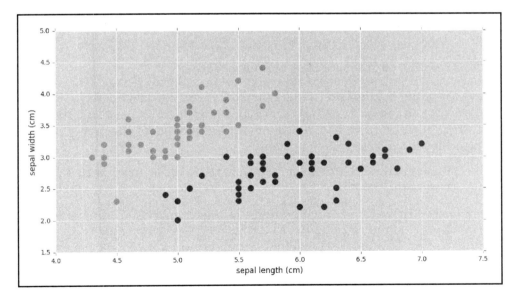

Plotting the first two features of the Iris dataset

Splitting the data into training and test sets

We learned in the previous chapter that it is essential to keep training and test data separate. We can easily split the data using one of scikit-learn's many helper functions:

```
In [11]: X_train, X_test, y_train, y_test =
model_selection.train_test_split(
...             data, target, test_size=0.1, random_state=42
...     )
```

Here we want to split the data into 90 percent training data and 10 percent test data, which we specify with `test_size=0.1`. By inspecting the return arguments, we note that we ended up with exactly 90 training data points and 10 test data points:

```
In [12]: X_train.shape, y_train.shape
Out[12]: ((90, 4), (90,))
In [13]: X_test.shape, y_test.shape
Out[13]: ((10, 4), (10,))
```

Training the classifier

Creating a logistic regression classifier involves pretty much the same steps as setting up k-NN:

```
In [14]: lr = cv2.ml.LogisticRegression_create()
```

We then have to specify the desired training method. Here, we can choose `cv2.ml.LogisticRegression_BATCH` or `cv2.ml.LogisticRegression_MINI_BATCH`. For now, all we need to know is that we want to update the model after every data point, which can be achieved with the following code:

```
In [15]: lr.setTrainMethod(cv2.ml.LogisticRegression_MINI_BATCH)
...      lr.setMiniBatchSize(1)
```

We also want to specify the number of iterations the algorithm should run before it terminates:

```
In [16]: lr.setIterations(100)
```

We can then call the `train` method of the object (in the exact same way as we did earlier), which will return `True` upon success:

```
In [17]: lr.train(X_train, cv2.ml.ROW_SAMPLE, y_train)
Out[17]: True
```

As we just saw, the goal of the training phase is to find a set of weights that best transform the feature values into an output label. A single data point is given by its four feature values (f_0, f_1, f_2, f_3). Since we have four features, we should also get four weights, so that $x = w_0 f_0 + w_1 f_1 + w_2 f_2 + w_3 f_3$, and $\hat{y}=\sigma(x)$. However, as discussed previously, the algorithm adds an extra weight that acts as an offset or bias, so that $x = w_0 f_0 + w_1 f_1 + w_2 f_2 + w_3 f_3 + w_4$. We can retrieve these weights as follows:

```
In [18]: lr.get_learnt_thetas()
Out[18]: array([[-0.04109113, -0.01968078, -0.16216497, 0.28704911,
0.11945518]], dtype=float32)
```

This means that the input to the logistic function is $x = -0.0411 f_0 - 0.0197 f_1 - 0.162 f_2 + 0.287 f_3 + 0.119$. Then, when we feed in a new data point (f_0, f_1, f_2, f_3) that belongs to class 1, the output $\hat{y}=\sigma(x)$ should be close to 1. But how well does that actually work?

Testing the classifier

Let's see for ourselves by calculating the accuracy score on the training set:

```
In [19]: ret, y_pred = lr.predict(X_train)
In [20]: metrics.accuracy_score(y_train, y_pred)
Out[20]: 1.0
```

Perfect score! However, this only means that the model was able to perfectly **memorize** the training dataset. This does not mean that the model would be able to classify a new, unseen data point. For this, we need to check the test dataset:

```
In [21]: ret, y_pred = lr.predict(X_test)
    ...      metrics.accuracy_score(y_test, y_pred)
Out[21]: 1.0
```

Luckily, we get another perfect score! Now we can be sure that the model we built is truly awesome.

Summary

In this chapter, we covered quite a lot of ground, didn't we!

In short, we learned a lot about different supervised learning algorithms, how to apply them to real datasets, and how to implement everything in OpenCV. We introduced classification algorithms such as k-NN and logistic regression and discussed how they could be used to predict labels as two or more discrete categories. We introduced various variants of linear regression (such as Lasso regression and ridge regression) and discussed how they could be used to predict continuous variables. Last but not least, we got acquainted with the Iris and Boston datasets, two classics in the history of machine learning.

In the following chapters, we will go into much greater depth within these topics, and see some more interesting examples of where these concepts can be useful.

But first, we need to talk about another essential topic of machine learning, **feature engineering**. Often, data does not come in nicely formatted datasets, and it is our responsibility to represent the data in a meaningful way. Therefore, the next chapter will talk all about representing features and engineering data.

4
Representing Data and Engineering Features

In the last chapter, we built our very first supervised learning models and applied them to some classic datasets, such as the **Iris** and the **Boston** datasets. However, in the real world, data rarely comes in a neat `<n_samples x n_features>` **feature matrix** that is part of a pre-packaged database. Instead, it is our own responsibility to find a way to represent the data in a meaningful way. The process of finding the best way to represent our data is known as **feature engineering**, and it is one of the main tasks of data scientists and machine learning practitioners trying to solve real-world problems.

I know you would rather jump right to the end and build the deepest neural network mankind has ever seen. But, trust me, this stuff is important! Representing our data in the right way can have a much greater influence on the performance of our supervised model than the exact parameters we choose. And we get to invent our own features, too.

In this chapter, we will therefore go over some common feature engineering tasks. Specifically, we want to answer the following questions:

- What are some common preprocessing techniques that everyone uses but nobody talks about?
- How do we represent categorical variables, such as the names of products, of colors, or of fruits?
- How would we even go about representing text?
- What is the best way to encode images, and what do SIFT and SURF stand for?

Let's start from the top.

Understanding feature engineering

Believe it or not, how well a machine learning system can learn is mainly determined by the quality of the training data. Although every learning algorithm has its own strengths and weaknesses, differences in performance often come down to the way the data is prepared or represented. Feature engineering can thus be understood as a tool for **data representation**. Machine learning algorithms try to learn a solution to a problem from sample data, and feature engineering asks: what is the best representation of the sample data to learn a solution to the problem?

Remember, a couple of chapters ago, we talked about the whole machine learning pipeline. There we already mentioned feature extraction, but we didn't really discuss what it is all about. Let's see how it fits in the pipeline:

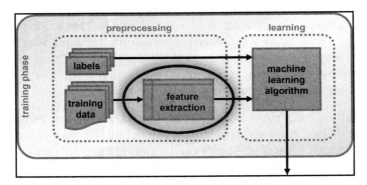

Feature engineering is part of the data preprocessing pipeline

As a brief reminder, we have already discussed that feature engineering can be split into two stages:

- **Feature selection**: This is the process of identifying important attributes (or features) in the data. Possible features of an image might be the location of edges, corners, or ridges. In this chapter, we will look at some of the more advanced feature descriptors that OpenCV provides, such as the **Scale-Invariant Feature Transform (SIFT)** and **Speeded Up Robust Features (SURF)**.

- **Feature extraction**: This is the actual process of transforming the raw data into the desired **feature space** used to feed a machine learning algorithm, as illustrated in the figure shown earlier. An example would be the **Harris operator**, which allows us to extract corners (that is, a selected feature) in an image.

> I often hear people say, *feature engineering is an art*. Although I believe this to be true for the masters of the field, who truly have an eye for the best feature transform, I can't help but think of this as a loaded term, most often used to mystify the topic of feature engineering and, who knows, maybe sell some books on the side. At the end of the day, there are some well-established ways to preprocess data, which have only little to do with art.

What's left to do is to walk through these procedures step-by-step and discuss some of the most common data preprocessing techniques.

Preprocessing data

The more disciplined we are in handling our data, the better results we are likely to achieve in the end. The first step in this procedure is known as **data preprocessing**, and it comes in (at least) three different flavors:

- **Data formatting**: The data may not be in a format that is suitable for us to work with. For example, the data might be provided in a proprietary file format, which our favorite machine learning algorithm does not understand.
- **Data cleaning**: The data may contain invalid or missing entries, which need to be cleaned up or removed.
- **Data sampling**: The data may be far too large for our specific purpose, forcing us to sample the data in a smart way.

Once the data has been preprocessed, we are ready for the actual feature engineering: to **transform the preprocessed data** to fit our specific machine learning algorithm. This step usually involves one or more of three possible procedures:

- **Scaling**: Certain machine learning algorithms often require the data to be within a common range, such as to have zero mean and unit variance. Scaling is the process of bringing all features (which might have different physical units) into a common range of values.

- **Decomposition**: Datasets often have many more features than we could possibly process. Feature decomposition is the process of compressing data into a smaller number of highly informative data components.
- **Aggregation**: Sometimes, it is possible to group multiple features into a single, more meaningful one. For example, a database might contain the date and time for each user who logged into a web-based system. Depending on the task, this data might be better represented by simply counting the number of logins per user.

Let's look at some of these processes in more detail.

Standardizing features

Standardization refers to the process of scaling the data to have zero mean and unit variance. This is a common requirement for a wide range of machine learning algorithms, which might behave badly if individual features do not fulfill this requirement. We could manually standardize our data by subtracting from every data point the mean value (μ) of all the data, and dividing that by the variance (σ) of the data; that is, for every feature x, we would compute $(x - \mu) / \sigma$.

Alternatively, scikit-learn offers a straightforward implementation of this process in its `preprocessing` module. Let's consider a 3 x 3 data matrix X, standing for three data points (rows) with three arbitrarily chosen feature values each (columns):

```
In [1]: from sklearn import preprocessing
   ...: import numpy as np
   ...: X = np.array([[ 1., -2., 2.],
   ...:               [ 3.,  0., 0.],
   ...:               [ 0.,  1., -1.]])
```

Then, standardizing the data matrix X can be achieved with the function `scale`:

```
In [2]: X_scaled = preprocessing.scale(X)
   ...: X_scaled
Out[2]: array([[-0.26726124, -1.33630621,  1.33630621],
               [ 1.33630621,  0.26726124, -0.26726124],
               [-1.06904497,  1.06904497, -1.06904497]])
```

We can verify that the scaled data matrix, `X_scaled`, is indeed standardized by double-checking the mean and variance. A standardized feature matrix should have a mean value of (close to) zero in every row:

```
In [3]: X_scaled.mean(axis=0)
Out[3]: array([ 7.40148683e-17, 0.00000000e+00, 0.00000000e+00])
```

In addition, every row of the standardized feature matrix should have variance of 1 (which is the same as checking for a standard deviation of 1 using `std`):

```
In [4]: X_scaled.std(axis=0)
Out[4]: array([ 1., 1., 1.])
```

Normalizing features

Similar to standardization, **normalization** is the process of scaling individual samples to have unit norm. I'm sure you know that the norm stands for the **length of a vector**, and can be defined in different ways. We discussed two of them in the previous chapter: the **L1 norm** (or **Manhattan distance**) and the **L2 norm** (or **Euclidean distance**).

In scikit-learn, our data matrix `X` can be normalized using the `normalize` function, and the `l1` norm is specified by the `norm` keyword:

```
In [5]: X_normalized_l1 = preprocessing.normalize(X, norm='l1')
   ...: X_normalized_l1
Out[5]: array([[ 0.2, -0.4, 0.4],
               [ 1. ,  0. , 0. ],
               [ 0. ,  0.5, -0.5]])
```

Similarly, the L2 norm can be computed by specifying `norm='l2'`:

```
In [6]: X_normalized_l2 = preprocessing.normalize(X, norm='l2')
   ...: X_normalized_l2
Out[6]: array([[ 0.33333333, -0.66666667,  0.66666667],
               [ 1.        ,  0.        ,  0.        ],
               [ 0.        ,  0.70710678, -0.70710678]])
```

Scaling features to a range

An alternative to scaling features to zero mean and unit variance is to get features to lie between a given minimum and maximum value. Often these values are zero and one, so that the maximum absolute value of each feature is scaled to unit size. In scikit-learn, this can be achieved using `MinMaxScaler`:

```
In [7]: min_max_scaler = preprocessing.MinMaxScaler()
   ...: X_min_max = min_max_scaler.fit_transform(X)
   ...: X_min_max
Out[7]: array([[ 0.33333333, 0. , 1. ],
               [ 1. , 0.66666667, 0.33333333],
               [ 0. , 1. , 0. ]])
```

By default, the data will be scaled to fall within 0 and 1. We can specify different ranges by passing a keyword argument `feature_range` to the `MinMaxScaler` constructor:

```
In [8]: min_max_scaler = preprocessing.MinMaxScaler(feature_range
   ...:                                             (-10,10))
   ...: X_min_max2 = min_max_scaler.fit_transform(X)
   ...: X_min_max2
Out[8]: array([[ -3.33333333, -10. , 10. ],
               [ 10. , 3.33333333, -3.33333333],
               [-10. , 10. , -10. ]])
```

Binarizing features

Finally, we might find ourselves not caring too much about the exact feature values of the data. Instead, we might just want to know if a feature is present or absent. **Binarizing** the data can be achieved by **thresholding** the feature values. Let's quickly remind ourselves of our feature matrix, X:

```
In [9]: X
Out[9]: array([[ 1., -2., 2.],
               [ 3., 0., 0.],
               [ 0., 1., -1.]])
```

Let's assume that these numbers represent the thousands of dollars in our bank accounts. If there are more than 0.5 thousand dollars in the account, we consider the person rich, which we represent with a 1. Else we put a 0. This is akin to thresholding the data with `threshold=0.5`:

```
In [10]: binarizer = preprocessing.Binarizer(threshold=0.5)
    ...     X_binarized = binarizer.transform(X)
    ...     X_binarized
Out[10]: array([[ 1.,  0.,  1.],
                [ 1.,  0.,  0.],
                [ 0.,  1.,  0.]])
```

The result is a matrix made entirely of ones and zeros.

Handling the missing data

Another common need in feature engineering is the handling of missing data. For example, we might have a dataset that looks like this:

```
In [11]: from numpy import nan
    ...     X = np.array([[ nan,  0,   3 ],
    ...                   [ 2,    9,  -8 ],
    ...                   [ 1,   nan,  1 ],
    ...                   [ 5,    2,   4 ],
    ...                   [ 7,    6,  -3 ]])
```

Most machine learning algorithms cannot handle the **Not a Number** (**NAN**) values (nan in Python). Instead, we first have to replace all the nan values with some appropriate fill values. This is known as **imputation** of missing values.

Three different strategies to impute missing values are offered by scikit-learn:

- `'mean'`: Replaces all the nan values with mean value along a specified axis of the matrix (default: *axis=0*)
- `'median'`: Replaces all the nan values with median value along a specified axis of the matrix (default: *axis=0*)
- `'most_frequent'`: Replaces all the nan values with the most frequent value along a specified axis of the matrix (default: *axis=0*)

For example, the `'mean'` imputer can be called as follows:

```
In [12]: from sklearn.preprocessing import Imputer
   ...:  imp = Imputer(strategy='mean')
   ...:  X2 = imp.fit_transform(X)
   ...:  X2
Out[12]: array([[ 3.75, 0.  ,  3. ],
                [ 2.  , 9.  , -8. ],
                [ 1.  , 4.25,  1. ],
                [ 5.  , 2.  ,  4. ],
                [ 7.  , 6.  , -3. ]])
```

This replaced the two `nan` values with fill values equivalent to the mean value calculated along the corresponding columns. We can double-check the math by calculating the mean across the first column (without counting the first element `X[0, 0]`), and comparing this number to the first element in the matrix, `X2[0, 0]`:

```
In [13]: np.mean(X[1:, 0]), X2[0, 0]
Out[13]: (3.75, 3.75)
```

Similarly, the `'median'` strategy relies on the same code:

```
In [14]: imp = Imputer(strategy='median')
   ...:  X3 = imp.fit_transform(X)
   ...:  X3
Out[14]: array([[ 3.5, 0. ,  3. ],
                [ 2. , 9. , -8. ],
                [ 1. , 4. ,  1. ],
                [ 5. , 2. ,  4. ],
                [ 7. , 6. , -3. ]])
```

Let's double-check the math one more time. This time we won't calculate the mean of the first column but the median (without including `X[0, 0]`), and compare the result to `X3[0, 0]`. We find that the two values are the same, convincing us that `Imputer` works as expected:

```
In [15]: np.median(X[1:, 0]), X3[0, 0]
Out[15]: (3.5, 3.5)
```

Understanding dimensionality reduction

Datasets often have many more features than we could possibly process. For example, let's say our job was to predict a country's poverty rate. We would probably start by matching a country's name with its poverty rate, but that would not help us predict the poverty rate of a new country. So we start thinking about possible causes of poverty. But how many possible causes of poverty are there? Factors might include a country's economy, lack of education, high divorce rate, overpopulation, and so on. If each one of these causes was a feature used to help predict the poverty rate, we would end up with a countless number of features. If you're a mathematician, you might think of these features as **axes** in a **high-dimensional space**, and every country's poverty rate is then a single point in this high-dimensional space.

If you're not a mathematician, it might help to start small. Let's say, we first look at only two features: a country's **Gross Domestic Product** (**GDP**) and the number of citizens. We interpret the GDP as *x-axis* and the number of citizens as *y-axis* in a 2D space. Then we look at the first country. It has a small GDP and an average number of citizens. We draw a point in the *x-y* plane that represents this country. We add a second, a third, and a fourth country. The fourth country just happens to have both a high GDP and a large number of citizens. Hence, our four data points might be spread across the *x-y* plane much like in the following figure:

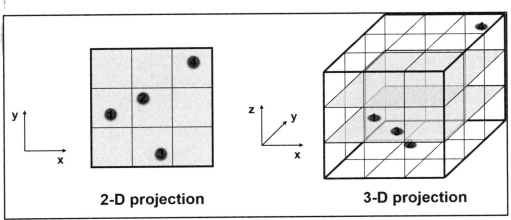

Projecting data points into a 2D and 3D subspace

Representing Data and Engineering Features

However, what happens if we start adding a third feature, such as the country's divorce rate, to our analysis? This would add a third axis to our plot (*z axis*). Suddenly, we find that the data no longer spreads very nicely across the *x-y-z* cube, as most of the cube remains empty. While in two dimensions it seemed like we had most of the *x-y* square covered, in three dimensions we would need many more data points to fill the void between the data points 1 to 3 and the lonely data point 4 in the upper-right corner.

This problem is also known as the **curse of dimensionality**: the number of data points needed to fill the available space grows exponentially with the number of dimensions (or plot axes). If a classifier is not fed with data points that span the entire feature space (such as shown in the preceding cube example), the classifier will not know what to do once a new data point is presented that lies far away from all the previously encountered data points.

In practice, the curse of dimensionality means that for a given sample size, there is a maximum number of features, above which the performance of our classifier will degrade rather than improve:

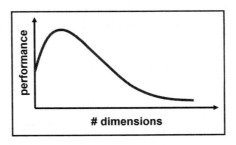

(Hypothetical) performance as a function of data dimensionality

But, how do we find this seemingly optimal number of dimensions for our dataset?

This is where **dimensionality reduction** comes in to play. These are a family of techniques that allow us to find a compact representation of our high-dimensional data, without losing too much information.

Implementing Principal Component Analysis (PCA) in OpenCV

One of the most common dimensionality reduction techniques is called **Principal Component Analysis (PCA)**.

Similar to the 2D and 3D examples shown earlier, we can think of an image as a point in a high-dimensional space. If we flatten a 2D image of height m and width n by stacking all the columns, we get a (feature) vector of length $m\,n \times 1$. The value of the i-th element in this vector is the grayscale value of the i-th pixel in the image. Now, imagine we would like to represent every possible 2D grayscale image with these exact dimensions. How many images would that give?

Since grayscale pixels usually take values between 0 and 255, there are a total of 256 raised to the power of $m\,n$ images. Chances are that this number is huge! So, naturally, we ask ourselves if there could be a smaller, more compact representation (using less than $m\,n$ features) that describes all these images equally well. After all, we have previously realized that grayscale values are not the most informative measures of content. So maybe there's a better way of representing all possible images than simply looking at all their grayscale values.

This is where PCA comes in. Consider a dataset from which we extract exactly two features. These features could be the grayscale values of pixels at some x and y positions, but they could also be more complex than that. If we plot the dataset along these two feature axes, the data might lie within some **multivariate Gaussian**:

```
In [1]: import numpy as np
   ...      mean = [20, 20]
   ...      cov = [[5, 0], [25, 25]]
   ...      x, y = np.random.multivariate_normal(mean, cov, 1000).T
```

We can plot this data using Matplotlib:

```
In [2]: import matplotlib.pyplot as plt
   ...      plt.style.use('ggplot')
   ...      %matplotlib inline
In [3]: plt.plot(x, y, 'o', zorder=1)
   ...      plt.axis([0, 40, 0, 40])
   ...      plt.xlabel('feature 1')
   ...      plt.ylabel('feature 2')
Out[3]: <matplotlib.text.Text at 0x125e0c9af60>
```

Representing Data and Engineering Features

This will produce the following plot:

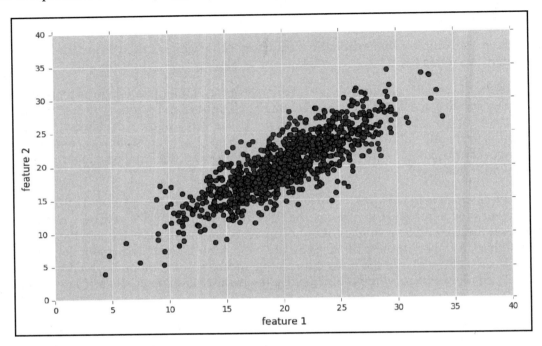

Data drawn from a multivariate Gaussian

What PCA does is rotate all the data points until the data lies aligned with the two axes that explain most of the spread of the data. Let's have a look at what that means.

In OpenCV, performing PCA is as simple as calling `cv2.PCACompute`. However, first we have to stack the feature vectors x and y into a single feature matrix, X:

```
In [2]: X = np.vstack((x, y)).T
```

Then we can compute PCA on the feature matrix X. We also specify an empty array, `np.array([])`, for the mask argument, which tells OpenCV to use all data points in the feature matrix:

```
In [3]: import cv2
   ...: mu, eig = cv2.PCACompute(X, np.array([]))
   ...: eig
Out[3]: array([[ 0.71160978,  0.70257493],
               [-0.70257493,  0.71160978]])
```

The function returns two values: the mean value subtracted before the projection (`mean`), and the eigenvectors of the covariation matrix (`eig`). These eigenvectors point to the direction PCA considers the most informative. If we plot them on top of our data using Matplotlib, we find that they are aligned with the spread of the data:

```
In [5]: plt.plot(x, y, 'o', zorder=1)
   ...      plt.quiver(mean[0], mean[1], eig[:, 0], eig[:, 1], zorder=3,
scale=0.2,
   ...                 units='xy')
```

We also add some text labels to the eigenvectors:

```
   ...      plt.text(mean[0] + 5 * eig[0, 0], mean[1] + 5 * eig[0, 1],
            'u1', zorder=5, fontsize=16, bbox=dict(facecolor='white',
            alpha=0.6))
   ...      plt.text(mean[0] + 7 * eig[1, 0], mean[1] + 4 * eig[1, 1],
            'u2', zorder=5, fontsize=16, bbox=dict(facecolor='white',
            alpha=0.6))
   ...      plt.axis([0, 40, 0, 40])
   ...      plt.xlabel('feature 1')
   ...      plt.ylabel('feature 2')
Out[7]: <matplotlib.text.Text at 0x1f3499f5860>
```

This will produce the following figure:

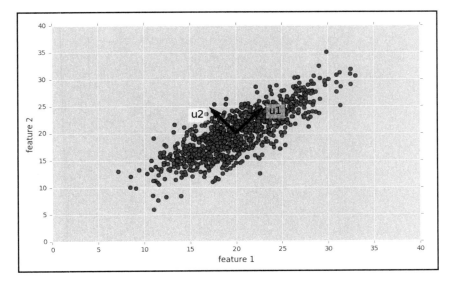

Principal Component Analysis (PCA)

Interestingly, the first eigenvector (labeled **u1** in the preceding figure) points in the direction where the spread of the data is maximal. This is called the first **principal component** of the data. The second principal component is **u2**, and indicates the axis along the second most variation in the data can be observed.

Thus, what PCA is telling us is that our predetermined x and y axes were not really that meaningful to describe the data we had chosen. Because the spread of the chosen data is angled at roughly 45 degrees, it would make more sense to chose **u1** and **u2** as axes instead of x and y.

To prove this point, we can rotate the data by using cv2.PCAProject:

```
In [16]: X2 = cv2.PCAProject(X, mu, eig)
```

By doing this, the data should be rotated so that the two axes of maximal spread are aligned with the x and y axes. We can convince ourselves of this fact by plotting the new data matrix, X2:

```
In [17]: plt.plot(X2[:, 0], X2[:, 1], 'o')
   ...:  plt.xlabel('first principal component')
   ...:  plt.ylabel('second principal component')
   ...:  plt.axis([-20, 20, -10, 10])
```

This leads to the following plot:

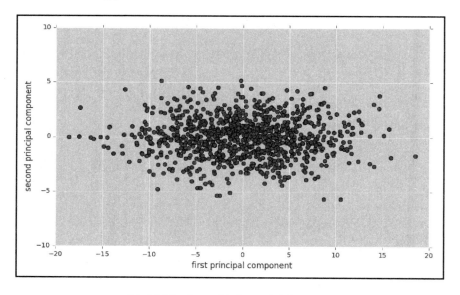

Data plotted along the axes of the first two principal components

In this image, we can see that the data is spread out the most along the *x*-axis. Hence, we are convinced that the projection was successful.

Implementing Independent Component Analysis (ICA)

Other useful dimensionality reduction techniques that are closely related to PCA are provided by scikit-learn, but not OpenCV. We mention them here for the sake of completeness. **Independent Component Analysis (ICA)** performs the same mathematical steps as PCA, but it chooses the components of the decomposition to be as independent as possible from each other.

In scikit-learn, ICA is available from the `decomposition` module:

```
In [8]: from sklearn import decomposition
In [9]: ica = decomposition.FastICA()
```

As seen before, the data transformation happens in the function `fit_transform`:

```
In [10]: X2 = ica.fit_transform(X)
```

In our case, plotting the rotated data leads to a similar result as achieved with PCA earlier:

```
In [11]: plt.plot(X2[:, 0], X2[:, 1], 'o')
   ...:  plt.xlabel('first independent component')
   ...:  plt.ylabel('second independent component')
   ...:  plt.axis([-0.2, 0.2, -0.2, 0.2])
Out[11]: [-0.2, 0.2, -0.2, 0.2]
```

This can be seen in the following plot:

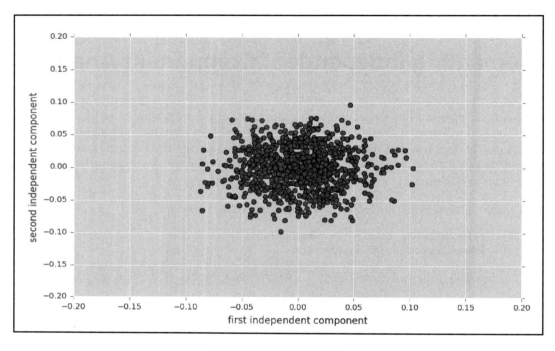

Independent Component Analysis (ICA)

Implementing Non-negative Matrix Factorization (NMF)

Another useful dimensionality reduction technique is called **Non-negative Matrix Factorization** (**NMF**). It again implements the same basic mathematical operations as PCA and ICA, but it has the additional constraint that it only operates on non-negative data. In other words, we cannot have negative values in our feature matrix if we want to use NMF; the resulting components of the decomposition will all have non-negative values as well.

In scikit-learn, NMF works exactly like ICA:

```
In [12]: nmf = decomposition.NMF()
In [13]: X2 = nmf.fit_transform(X)
In [14]: plt.plot(X2[:, 0], X2[:, 1], 'o')
    ...      plt.xlabel('first non-negative component')
    ...      plt.ylabel('second non-negative component')
    ...      plt.axis([-5, 15, -5, 15])
Out[14]: [-5, 15, -5, 15]
```

However, the resulting decomposition looks noticeably different from both PCA and ICA, which is a hallmark of NMF:

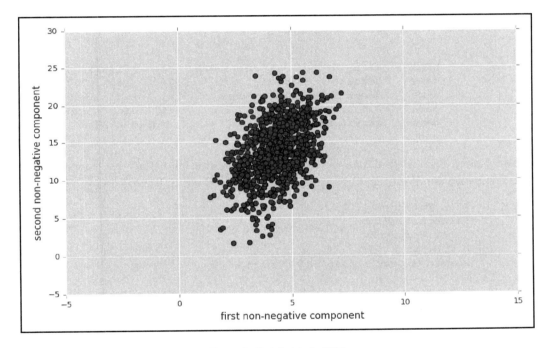

Non-negative Matrix Factorization (NMF)

Now that we are familiar with some of the most common data decomposition tools, let's have a look at some common data types.

Representing categorical variables

One of the most common data types we might encounter while building a machine learning system are **categorical features** (also known as **discrete features**), such as the color of a fruit or the name of a company. The challenge with categorical features is that they don't change in a continuous way, which makes it hard to represent them with numbers. For example, a banana is either green or yellow, but not both. A product belongs either in the clothing department or in the books department, but rarely in both, and so on.

How would you go about representing such features?

For example, let's assume we are trying to encode a dataset consisting of a list of forefathers of machine learning and artificial intelligence:

```
In [1]: data = [
   ...      {'name': 'Alan Turing', 'born': 1912, 'died': 1954},
   ...      {'name': 'Herbert A. Simon', 'born': 1916, 'died': 2001},
   ...      {'name': 'Jacek Karpinski', 'born': 1927, 'died': 2010},
   ...      {'name': 'J.C.R. Licklider', 'born': 1915, 'died': 1990},
   ...      {'name': 'Marvin Minsky', 'born': 1927, 'died': 2016}
   ...  ]
```

While the features `'born'` and `'died'` are already in numeric format, the `'name'` feature is a bit trickier to encode. We might be intrigued to encode them in the following way:

```
In [2]: {'Alan Turing': 1,
   ...   'Herbert A. Simon': 2,
   ...   'Jacek Karpinsky': 3,
   ...   'J.C.R. Licklider': 4,
   ...   'Marvin Minsky': 5};
```

Although this seems like a good idea, it does not make much sense from a machine learning perspective. Why not? Well, by assigning ordinal values to these categories, most machine learning algorithms go on to think that `'Alan Turing'` < `'Herbert A. Simon'` < `'Jacek Karpsinky'`, since 1 < 2 < 3. This is clearly not what we meant to say.

Instead, what we really meant to say was something along the lines of: the first data point belongs to the category `'Alan Turing'`, and it does not belong to category `'Herbert A. Simon'`, and not to category `'Jacek Karpsinky'`, and so on. In other words, we were looking for a **binary encoding**. In machine learning lingo, this is known as a **one-hot** encoding, and is provided by most machine learning packages straight out-of-the-box (except for OpenCV, of course).

In scikit-learn, one-hot encoding is provided by the class `DictVectorizer`, which can be found in the `feature_extraction` module. The way it works is by simply feeding a dictionary containing the data to the `fit_transform` function, and the function automatically determines which features to encode:

```
In [3]: from sklearn.feature_extraction import DictVectorizer
   ...: vec = DictVectorizer(sparse=False, dtype=int)
   ...: vec.fit_transform(data)
Out[3]: array([[1912, 1954, 1, 0, 0, 0, 0],
               [1916, 2001, 0, 1, 0, 0, 0],
               [1927, 2010, 0, 0, 0, 1, 0],
               [1915, 1990, 0, 0, 1, 0, 0],
               [1927, 2016, 0, 0, 0, 0, 1]], dtype=int32)
```

What happened here? The two year entries are still intact, but the rest of the rows have been replaced by ones and zeros. We can call `get_feature_names` to find out the listed order of the features:

```
In [4]: vec.get_feature_names()
Out[4]: ['born',
         'died',
         'name=Alan Turing',
         'name=Herbert A. Simon',
         'name=J.C.R. Licklider',
         'name=Jacek Karpinski',
         'name=Marvin Minsky']
```

The first row of our data matrix, which stands for Alan Turing, is now encoded as `'born'`=1912, `'died'`=1954, `'Alan Turing'`=1, `'Herbert A. Simon'`=0, `'J.C.R Licklider'`=0, `'Jacek Karpinsik'`=0, and `'Marvin Minsky'`=0.

There is one problem with this approach though. If our feature category has a lot of possible values, such as every possible first and last name, then one-hot encoding will lead to a very large data matrix. However, if we investigate the data matrix row-by-row, it becomes clear that every row has exactly one 1 and all the other entries are 0. In other words, the matrix is **sparse**. The scikit-learn provides a compact representation of sparse matrices, which we can trigger by specifying `sparse=True` in the first call to `DictVectorizer`:

```
In [5]: vec = DictVectorizer(sparse=True, dtype=int)
   ...: vec.fit_transform(data)
Out[5]: <5x7 sparse matrix of type '<class 'numpy.int32'>'
         with 15 stored elements in Compressed Sparse Row format>
```

Certain machine learning algorithms, such as decision trees, are capable of handling categorical features natively. In these cases, it is not necessary to use one-hot encoding.

We will come back to this technique when we talk about neural networks in `Chapter 9, Using Deep Learning to Classify Handwritten Digits`.

Representing text features

Similar to categorical features, scikit-learn offers an easy way to encode another common feature type, **text features**. When working with text features, it is often convenient to encode individual words or phrases as numerical values.

Let's consider a dataset that contains a small corpus of text phrases:

```
In [1]: sample = [
   ...         'feature engineering',
   ...         'feature selection',
   ...         'feature extraction'
   ...         ]
```

One of the simplest methods of encoding such data is by **word count**; for each phrase, we simply count the occurrences of each word within it. In scikit-learn, this is easily done using `CountVectorizer`, which functions akin to `DictVectorizer`:

```
In [2]: from sklearn.feature_extraction.text import CountVectorizer
   ...     vec = CountVectorizer()
   ...     X = vec.fit_transform(sample)
   ...     X
Out[2]: <3x4 sparse matrix of type '<class 'numpy.int64'>'
             with 6 stored elements in Compressed Sparse Row format>
```

By default, this will store our feature matrix X as a sparse matrix. If we want to manually inspect it, we need to convert it to a regular array:

```
In [3]: X.toarray()
Out[3]: array([[1, 0, 1, 0],
               [0, 0, 1, 1],
               [0, 1, 1, 0]], dtype=int64)
```

In order to understand what these numbers mean, we have to look at the feature names:

```
In [4]: vec.get_feature_names()
Out[4]: ['engineering', 'extraction', 'feature', 'selection']
```

Now it becomes clear what the integers in X mean. If we look at the phrase that is represented in the top row of X, we see that it contains one occurrence of the word `'engineering'` and one occurrence of the word `'feature'`. On the other hand, it does not contain the words `'extraction'` or `'selection'`. Does this make sense? A quick glance at our original data `sample` reveals that the phrase was indeed `'feature engineering'`.

Looking only at the array X (no cheating!), can you guess what the last phrase in `sample` was?

One possible shortcoming of this approach is that we might put too much weight on words that appear very frequently. One approach to fix this is known as **term frequency-inverse document frequency** (**TF-IDF**). What TF-IDF does might be easier to understand than its name, which is basically to weigh the word counts by a measure of how often they appear in the entire dataset.

The syntax for TF-IDF is pretty much similar to the previous command:

```
In [5]: from sklearn.feature_extraction.text import TfidfVectorizer
   ...: vec = TfidfVectorizer()
   ...: X = vec.fit_transform(sample)
   ...: X.toarray()
Out[5]: array([[ 0.861037 , 0.        , 0.50854232, 0.        ],
               [ 0.        , 0.        , 0.50854232, 0.861037 ],
               [ 0.        , 0.861037 , 0.50854232, 0.        ]])
```

We note that the numbers are now smaller than before, with the third column taking the biggest hit. This makes sense, as the third column corresponds to the most frequent word across all three phrases, `'feature'`:

```
In [6]: vec.get_feature_names()
Out[6]: ['engineering', 'extraction', 'feature', 'selection']
```

If you're interested in the math behind TF-IDF, its Wikipedia article is actually a pretty good starting point (https://en.wikipedia.org/wiki/Tf-idf). For more information about its specific implementation in scikit-learn, have a look at the API documentation at http://scikit-learn.org/stable/modules/feature_extraction.html#tfidf-term-weighting.

Representing text features will become important in Chapter 7, *Implementing a Spam Filter with Bayesian Learning*.

Representing images

One of the most common and important data types for computer vision are, of course, images. The most straightforward way to represent images is probably by using the grayscale value of each pixel in the image. Usually, grayscale values are not very indicative of the data they describe. For example, if we saw a single pixel with grayscale value 128, could we tell what object this pixel belonged to? Probably not. Therefore, grayscale values are not very effective **image features**.

Using color spaces

Alternatively, we might find that colors contain some information that raw grayscale values cannot capture. Most often, images come in the conventional RGB color space, where every pixel in the image gets an intensity value for its apparent **redness (R)**, **greenness (G)**, and **blueness (B)**. However, OpenCV offers a whole range of other color spaces, such as **Hue Saturation Value (HSV)**, **Hue Saturation Lightness (HSL)**, and the Lab color space. Let's have a quick look at them.

Encoding images in RGB space

I am sure that you are already familiar with the RGB color space, which uses **additive mixing** of different shades of red, green, and blue to produce different composite colors. The RGB color space is useful in everyday life, because it covers a large part of the color space that the human eye can see. This is why color television sets or color computer monitors only need to care about producing mixtures of red, green, and blue light.

In OpenCV, RGB images are supported straight out-of-the-box. All you need to know, or need to be reminded of, is that color images are actually stored as BGR images in OpenCV; that is, the order of color channels is blue-green-red instead of red-green-blue. The reasons for this choice of format are mostly historical. OpenCV has stated that the choice of this format was due to its popularity among camera manufacturers and software providers when OpenCV was first created.

We can load a sample image in BGR format using `cv2.imread`:

```
In [1]: import cv2
   ...      import matplotlib.pyplot as plt
   ...      %matplotlib inline
In [2]: img_bgr = cv2.imread('data/lena.jpg')
```

If you have ever tried to display a BGR image using Matplotlib or similar libraries, you might have noticed a weird blue tint to the image. This is due to the fact that Matplotlib expects an RGB image. To achieve this, we have to permute the color channels using `cv2.cvtColor`:

```
In [3]: img_rgb = cv2.cvtColor(img_bgr, cv2.COLOR_BGR2RGB)
```

For comparison, we can plot both `img_bgr` and `img_rgb` next to each other:

```
In [4]: plt.figure(figsize=(12, 6))
   ...      plt.subplot(121)
   ...      plt.imshow(img_bgr)
   ...      plt.subplot(122)
   ...      plt.imshow(img_rgb)
Out[4]: <matplotlib.image.AxesImage at 0x20f6d043198>
```

This code will lead to the following two images:

Using Matplotlib to display a picture of Lena-as BGR image (left) and RGB image (right)

In case you were wondering, Lena is the name of a popular Playmate (Lena Soderberg), and her picture is one of the most used images in computer history. The reason is that back in the 1970s, people got tired of using the conventional stock of test images, and instead wanted something glossy to ensure good output dynamic range. Or so they claim! It is, of course, no coincidence that the picture of an attractive woman became so popular in a male-dominated field, but that's not for us to judge.

Encoding images in HSV and HLS space

However, ever since the RGB color space was created, people have realized that it is actually quite a poor representation of human vision. Therefore, researchers have developed two alternative representations. One of them is called **HSV**, which stands for **hue, saturation, and value,** and the other one is called **HLS**, which stands for **hue, lightness, and saturation**. You might have seen these color spaces in **color pickers** and common image editing software. In these color spaces, the **hue** of the color is captured by a single hue channel, the colorfulness is captured by a saturation channel, and the lightness or brightness is captured by a lightness or **value** channel.

In OpenCV, an RGB image can easily be converted to HSV color space using `cv2.cvtColor`:

```
In [5]: img_hsv = cv2.cvtColor(img_bgr, cv2.COLOR_BGR2HSV)
```

The same is true for the HLS color space. In fact, OpenCV provides a whole range of additional color spaces, which are available via `cv2.cvtColor`. All we need to do is to replace the color flag with one of the following:

- HLS (hue, lightness, saturation) using `cv2.COLOR_BGR2HLS`
- LAB (lightness, green-red, and blue-yellow) using `cv2.COLOR_BGR2LAB`
- YUV (overall luminance, blue-luminance, red-luminance) using `cv2.COLOR_BGR2YUV`

Detecting corners in images

One of the most straightforward features to find in an image are probably **corners**. OpenCV provides at least two different algorithms to find corners in an image:

- **Harris corner detection**: Knowing that edges are areas with high-intensity changes in all directions, Harris and Stephens came up with a fast way of finding such areas. This algorithm is implemented as `cv2.cornerHarris` in OpenCV.

- **Shi-Tomasi corner detection**: Shi and Tomasi have their own idea of what are good features to track, and they usually do better than Harris corner detection by finding the N strongest corners. This algorithm is implemented as `cv2.goodFeaturesToTrack` in OpenCV.

Harris corner detection works only on grayscale images, so we first want to convert our BGR image to grayscale:

```
In [9]: img_gray = cv2.cvtColor(img_bgr, cv2.COLOR_BGR2GRAY)
```

The algorithm then expects an input image, the pixel neighborhood size considered for corner detection (`blockSize`), an aperture parameter for the edge detection (`ksize`), and the so-called Harris detector-free parameter (`k`):

```
In [10]: corners = cv2.cornerHarris(img_gray, 2, 3, 0.04)
```

We can plot the resulting corner map using Matplotlib:

```
In [11]: plt.imshow(corners, cmap='gray')
Out[11]: <matplotlib.image.AxesImage at 0x1f3ea003b70>
```

This produces a grayscale image that looks like this:

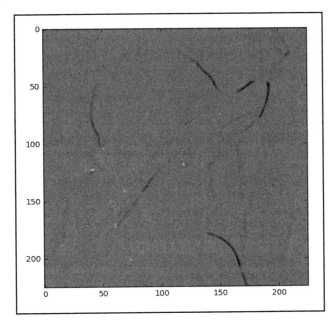

Harris corner detection

Using the Scale-Invariant Feature Transform (SIFT)

However, corner detection is not sufficient when the scale of an image changes. To this end, David Lowe came up with a method to describe interesting points in an image, independent of orientation and size. Hence, the name **scale-invariant feature transform (SIFT)**.

In OpenCV 3, this function is part of the `xfeatures2d` module:

```
In [6]: sift = cv2.xfeatures2d.SIFT_create()
```

The algorithm typically works in two steps:

- **Detect**: This step identifies interesting points in an image (also known as **keypoints**)
- **Compute**: This step computes the actual feature values for every keypoint

Keypoints can be detected with a single line of code:

```
In [7]: kp = sift.detect(img_bgr)
```

In addition, the `drawKeypoints` function offers a nice way to visualize the identified keypoints. By passing an optional flag, `cv2.DRAW_MATCHES_FLAGS_DRAW_RICH_KEYPOINTS`, the function surrounds every keypoint with a circle whose size denotes its importance, and a radial line that indicates the orientation of the keypoint:

```
In [8]: import numpy as np
   ...  img_kp = np.zeros_like(img_bgr)
   ...  img_kp = cv2.drawKeypoints(img_rgb, kp, img_kp,
        flags=cv2.DRAW_MATCHES_FLAGS_DRAW_RICH_KEYPOINTS)
In [9]: plt.imshow(img_kp)
Out[9]: <matplotlib.image.AxesImage at 0x1ed24f46550>
```

This will lead to the following plot:

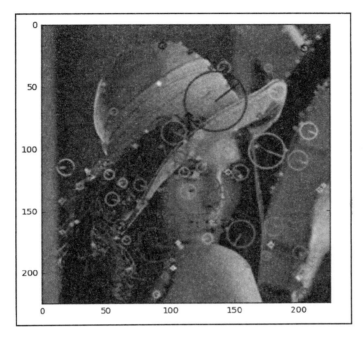

Applying SIFT to the Lena image

Feature descriptors can then be computed using the function `compute`:

```
In [10]: kp, des = sift.compute(img_bgr, kp)
```

The resulting feature descriptor (`des`) should have 128 feature values for every keypoint found. In our example, there seem to be a total of 238 features, each with 128 feature values:

```
In [11]: des.shape
Out[11]: (238, 128)
```

Alternatively, we can detect keypoints and compute feature descriptors in a single step:

```
kp2, des2 = sift.detectAndCompute(img_bgr, None)
```

Using NumPy, we can convince ourselves that both ways result in the same output, by making sure that every value in `des` is approximately the same as in `des2`:

```
In [13]: np.allclose(des, des2)
Out[13]: True
```

[109]

If you installed OpenCV 3 from source, the SIFT function might not be available. In such a case, you might have to obtain the extra modules from `https://github.com/Itseez/opencv_contrib` and reinstall OpenCV with the `OPENCV_EXTRA_MODULES_PATH` variable set. However, if you followed the installation instructions in Chapter 1, *A Taste of Machine Learning* and installed the Python Anaconda stack, you should have nothing to worry about.

Using Speeded Up Robust Features (SURF)

SIFT has proved to be really good, but it is not fast enough for most applications. This is where **Speeded Up Robust Features** (SURF) comes in, which has replaced the computationally expensive computations of SIFT with a box filter.

In OpenCV, SURF works in the exact same way as SIFT:

```
In [14]: surf = cv2.xfeatures2d.SURF_create()
In [15]: kp = surf.detect(img_bgr)
In [16]: img_kp = cv2.drawKeypoints(img_rgb, kp, img_kp,
         flags=cv2.DRAW_MATCHES_FLAGS_DRAW_RICH_KEYPOINTS)
In [17]: plt.imshow(img_kp)
Out[17]: <matplotlib.image.AxesImage at 0x2027dc494a8>
```

However, SURF detects a lot more features than SIFT, leading to the following plot:

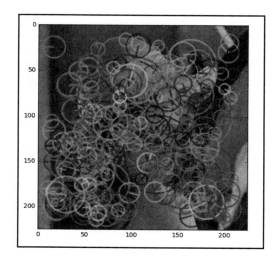

Applying SURF to the Lena image

After computing the feature descriptors, we can investigate the `des` object and realize that there are now 351 features found, each of which has 64 feature values:

```
In [18]: kp, des = surf.compute(img_bgr, kp)
In [19]: des.shape
Out[19]: (351, 64)
```

> Both SIFT and SURF are protected by patent laws. Therefore, if you wish to use them in a commercial application, you will be required to obtain a license. A free and open source alternative is `cv2.ORB`, which uses what is known as the **FAST keypoint detector** and the **BRIEF descriptor** to match SIFT and SURF in terms of performance and computational cost.

Summary

In this chapter, we went deep down the rabbit hole and looked at a number of common feature engineering techniques, focusing on both feature selection and feature extraction. We successfully formatted, cleaned, and transformed data so that it could be understood by common machine learning algorithms. We learned about the curse of dimensionality and dabbled a bit in dimensionality reduction by implementing PCA in OpenCV. Finally, we took a short tour of common feature extraction techniques that OpenCV provides for image data.

With these skills under our belt, we are now ready to take on any data, be it numerical, categorical, text, or image data. We know exactly what to do when we encounter missing data, and we know how to transfer our data to make it fit our preferred machine learning algorithm.

In the next chapter, we will take the next step and talk about a specific use-case, which is, how to use our newly gained knowledge to make medical diagnoses using decision trees.

5
Using Decision Trees to Make a Medical Diagnosis

Now that we know how to handle data in all shapes and forms, be it numerical, categorical, text, or image data, it is time to put our newly gained knowledge to good use.

In this chapter, we will learn how to build a machine learning system that can make a medical diagnosis. We aren't all doctors, but we've probably all been to one at some point in our lives. Typically, a doctor would gain as much information as possible about a patient's history and symptoms in order to make an informed diagnosis. We will mimic a doctor's decision-making process with the help of what is known as **decision trees**.

A decision tree is a simple yet powerful supervised learning algorithm that resembles a flow chart; we will talk more about this in just a minute. Other than in medicine, decision trees are commonly used in fields such as astronomy (for example, for filtering noise from Hubble Space Telescope images or to classify star-galaxy clusters), manufacturing and production (for example, by Boeing to discover flaws in the manufacturing process), and object recognition (for example, for recognizing 3D objects).

Specifically, we want to address the following questions in this chapter:

- How do we build simple decision trees from data, and use them for either classification or regression?
- How do we decide which decision to make next?
- How do we prune a tree, and what is that good for?

But first, let's talk about what decision trees actually are.

Understanding decision trees

A decision tree is simple yet powerful model for supervised learning problems. Like the name suggests, we can think of it as a tree in which information flows along different branches—starting at the trunk and going all the way to the individual leaves. If you are wondering if you have ever seen a decision tree before, let me remind you about the spam filter figure we encountered in Chapter 1, *A Taste of Machine Learning*:

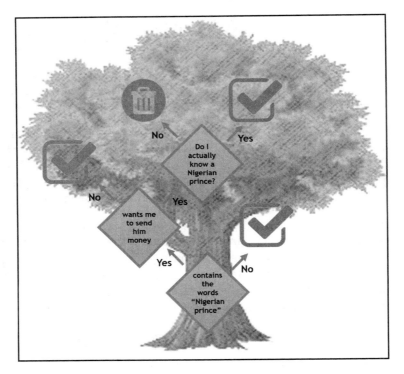

Example of a simple spam filter—revisited

This is basically a decision tree!

A decision tree is made of a hierarchy of questions or tests about the data (also known as **decision nodes**) and their possible consequences. In the preceding example, we might count the number of words in each email using the `CountVectorizer` object from the previous chapter. Then it's easy to ask the first question (whether the email contains the words **Nigerian prince**), simply by checking the count for the words **Nigerian** and **prince**. If these words are not part of the email, we would follow the edge that is labeled **No** in the preceding figure, and conclude that the email is harmless. Else, we would follow the edge that is labeled **Yes** in the preceding figure, and continue to test the email until it is safe to conclude that the message is spam.

From the preceding example, we may realize that one of the true difficulties with building decision trees is how to pull out suitable features from the data. To make this clear, let's be as concrete as possible. Let's say we have a dataset consisting of a single email:

```
In [1]: data = [
   ...       'I am Mohammed Abacha, the son of the late Nigerian Head of '
   ...       'State who died on the 8th of June 1998. Since i have been '
   ...       'unsuccessful in locating the relatives for over 2 years now '
   ...       'I seek your consent to present you as the next of kin so '
   ...       'that the proceeds of this account valued at US$15.5 Million '
   ...       'Dollars can be paid to you. If you are capable and willing '
   ...       'to assist, contact me at once via email with following '
   ...       'details: 1. Your full name, address, and telephone number. '
   ...       '2. Your Bank Name, Address. 3.Your Bank Account Number and '
   ...       'Beneficiary Name - You must be the signatory.'
   ...   ]
```

In case you were wondering, I have nothing against Nigerian princes or their email habits. But their reputation became victim of a popular email scam in the early 2000, which makes for a good pedagogic example. The previous email is a slightly shortened version of a common email scam listed on http://www.hoax-slayer.com/nigerian-scam-examples.html.

This email can be vectorized in much the same way as we did in the previous chapter, using scikit-learn's `CountVectorizer`:

```
In [2]: from sklearn.feature_extraction.text import CountVectorizer
   ...   vec = CountVectorizer()
   ...   X = vec.fit_transform(data)
```

From the previous chapter, we know that we can have a look at the feature names in X using the following function:

```
In [3]: function:vec.get_feature_names()[:5]
Out[3]: ['15', '1998', '8th', 'abacha', 'account']
```

For the sake of clarity, we focus on the first five words only, which are sorted alphabetically. Then the corresponding number of occurrences can be found as follows:

```
In [4]: X.toarray()[0, :5]
Out[4]: array([1, 1, 1, 1, 2], dtype=int64)
```

This tells us that four out of five words show up just once in the email, but the word `'account'` (the last one listed in `Out[3]`) actually shows up twice. Like in the last chapter, we typed `X.toarray()` in order to convert the sparse array X into a human-readable array. The result is a 2D array, where rows correspond to the data samples, and columns correspond to the feature names described in the preceding command. Since there is only one sample in the dataset, we limit ourselves to row 0 of the array (that is, the first data sample) and the first five columns in the array (that is, the first five words).

So, how do we check if the email is from a Nigerian prince?

One way to do this is to look if the email contains both the words `'nigerian'` and `'prince'`:

```
In [5]: 'nigerian' in vec.get_feature_names()
Out[5]: True
In [6]: 'prince' in vec.get_feature_names()
Out[6]: False
```

What do we find to our surprise? The word `'prince'` does not occur in the email.

Does this mean the message is legit?

No, of course not. Instead of `'prince'`, the email went with the words `'head of state'`, effectively circumventing our all-too-simple spam detector.

Similarly, how do we even start to model the second decision in the tree-*wants me to send him money*? There is no straightforward feature in the text that would answer this question. Hence, this is a problem of **feature engineering**, of combining the words that actually occur in the message in such a way that allows us to answer this question. Sure, a good sign would be to look for strings like `'US$'` and `'money'`, but then we still wouldn't know the context in which these words were mentioned. For all we know, perhaps they were part of the sentence: *Don't worry, I don't want you to send me any money.*

To make matters worse, it turns out that the order in which we ask these questions can actually influence the final outcome. For example, what if we asked the last question first: *do I actually know a Nigerian prince?* Suppose we had a Nigerian prince for an uncle, then finding the words *Nigerian prince* in the email might no longer be suspicious.

As you can see, this seemingly simple example got quickly out of hand.

Luckily, the theoretical framework behind decision trees provides us with help in finding both the right decision rules as well as which decisions to tackle next.

However, in order to understand these concepts, we have to dig a little deeper.

Building our first decision tree

I think we are ready for a more complex example. As promised earlier, let's move this into the medical domain.

Let's consider an example where a number of patients have suffered from the same illness, such as a rare form of **basorexia**. Let's further assume that the true causes of the disease remain unknown to this day, and that all the information that is available to us consists of a bunch of physiological measurements. For example, we might have access to the following information:

- a patient's blood pressure (`'BP'`)
- a patient's cholesterol level (`'cholesterol'`)
- a patient's gender (`'sex'`)
- a patient's age (`'age'`)
- a patient's blood sodium concentration (`'Na'`)
- a patient's blood potassium concentration (`'K'`)

Based on all this information, let's suppose a doctor made recommendations to his patient to treat their disease using one of four possible drugs, drug A, B, C, or D. We have data available for 20 different patients:

```
In [1]: data = [
   ...      {'age': 33, 'sex': 'F', 'BP': 'high', 'cholesterol': 'high',
   ...       'Na': 0.66, 'K': 0.06, 'drug': 'A'},
   ...      {'age': 77, 'sex': 'F', 'BP': 'high', 'cholesterol': 'normal',
   ...       'Na': 0.19, 'K': 0.03, 'drug': 'D'},
   ...      {'age': 88, 'sex': 'M', 'BP': 'normal', 'cholesterol': 'normal',
   ...       'Na': 0.80, 'K': 0.05, 'drug': 'B'},
   ...      {'age': 39, 'sex': 'F', 'BP': 'low', 'cholesterol': 'normal',
```

Using Decision Trees to Make a Medical Diagnosis

```
       ...       'Na': 0.19, 'K': 0.02, 'drug': 'C'},
       ...      {'age': 43, 'sex': 'M', 'BP': 'normal', 'cholesterol': 'high',
       ...       'Na': 0.36, 'K': 0.03, 'drug': 'D'},
       ...      {'age': 82, 'sex': 'F', 'BP': 'normal', 'cholesterol': 'normal',
       ...       'Na': 0.09, 'K': 0.09, 'drug': 'C'},
       ...      {'age': 40, 'sex': 'M', 'BP': 'high', 'cholesterol': 'normal',
       ...       'Na': 0.89, 'K': 0.02, 'drug': 'A'},
       ...      {'age': 88, 'sex': 'M', 'BP': 'normal', 'cholesterol': 'normal',
       ...       'Na': 0.80, 'K': 0.05, 'drug': 'B'},
       ...      {'age': 29, 'sex': 'F', 'BP': 'high', 'cholesterol': 'normal',
       ...       'Na': 0.35, 'K': 0.04, 'drug': 'D'},
       ...      {'age': 53, 'sex': 'F', 'BP': 'normal', 'cholesterol': 'normal',
       ...       'Na': 0.54, 'K': 0.06, 'drug': 'C'},
       ...      {'age': 63, 'sex': 'M', 'BP': 'low', 'cholesterol': 'high',
       ...       'Na': 0.86, 'K': 0.09, 'drug': 'B'},
       ...      {'age': 60, 'sex': 'M', 'BP': 'low', 'cholesterol': 'normal',
       ...       'Na': 0.66, 'K': 0.04, 'drug': 'C'},
       ...      {'age': 55, 'sex': 'M', 'BP': 'high', 'cholesterol': 'high',
       ...       'Na': 0.82, 'K': 0.04, 'drug': 'B'},
       ...      {'age': 35, 'sex': 'F', 'BP': 'normal', 'cholesterol': 'high',
       ...       'Na': 0.27, 'K': 0.03, 'drug': 'D'},
       ...      {'age': 23, 'sex': 'F', 'BP': 'high', 'cholesterol': 'high',
       ...       'Na': 0.55, 'K': 0.08, 'drug': 'A'},
       ...      {'age': 49, 'sex': 'F', 'BP': 'low', 'cholesterol': 'normal',
       ...       'Na': 0.27, 'K': 0.05, 'drug': 'C'},
       ...      {'age': 27, 'sex': 'M', 'BP': 'normal', 'cholesterol': 'normal',
       ...       'Na': 0.77, 'K': 0.02, 'drug': 'B'},
       ...      {'age': 51, 'sex': 'F', 'BP': 'low', 'cholesterol': 'high',
       ...       'Na': 0.20, 'K': 0.02, 'drug': 'D'},
       ...      {'age': 38, 'sex': 'M', 'BP': 'high', 'cholesterol': 'normal',
       ...       'Na': 0.78, 'K': 0.05, 'drug': 'A'}
       ...     ]
```

Here, the dataset is actually a list of dictionaries, where every dictionary constitutes a single data point that contains a patient's blood work, age, gender, as well as the drug that was prescribed.

What was the doctor's reasoning for prescribing drugs A, B, C, or D? Can we see a relationship between a patient's blood values and the drug that the doctor prescribed?

Chances are, this is as hard for you to see as it is for me, even though I invented the dataset myself! Although the dataset might look random at first glance, I have in fact put in some clear relationships between a patient's blood values and the prescribed drug. Let's see if a decision tree can uncover these hidden relationships.

Understanding the task by understanding the data

What is always the first step in tackling a new machine learning problem?

You are absolutely right: to get a sense of the data. The better we understand the data, the better we understand the problem we are trying to solve. In our future endeavors, this will also help us choose an appropriate machine learning algorithm.

The first thing to realize is that the `'drug'` column is actually not a feature value like all the other columns. Since it is our goal to predict which drug will be prescribed based on a patient's blood values, the `'drug'` column effectively becomes the **target labels**. In other words, the inputs to our machine learning algorithm will be all blood values, age, and gender of a patient. Thus, the output will be a prediction of which drug to prescribe. Since the `'drug'` column is categorical in nature and not numerical, we know that we are facing a **classification task**.

Thus, it would be a good idea to remove all `'drug'` entries from the dictionaries listed in the `'data'` variable and store them in a separate variable. For this, we need to go through the list and extract the `'drug'` entry, which is easiest to do with a **list comprehension**:

```
In [2]: target = [d['drug'] for d in data]
   ...  target
Out[2]: ['A', 'D', 'B', 'C', 'D', 'C', 'A', 'B', 'D', 'C',
         'A', 'B', 'C', 'B', 'D', 'A', 'C', 'B', 'D', 'A']
```

Since we also want to remove the `'drug'` entries in all the dictionaries, we need to go through the list again and pop the `'drug'` key. We add a ; to suppress the output, since we don't want to see the whole dataset again:

```
In [3]: [d.pop('drug') for d in data];
```

Sweet! Now let's look at the data. For the sake of simplicity, we may want to focus on the numerical features first: `'age'`, `'K'`, and `'Na'`. These are relatively easy to plot using Matplotlib's `scatter` function. We first import Matplotlib as usual:

```
In [4]: import matplotlib.pyplot as plt
   ...  %matplotlib inline
   ...  plt.style.use('ggplot')
```

Then, if we want to plot the potassium level against the sodium level for every data point in
the dataset, we need to go through the list and extract these feature values:

```
In [5]: age = [d['age'] for d in data]
   ...  age
Out[5]: [33, 77, 88, 39, 43, 82, 40, 88, 29, 53,
         36, 63, 60, 55, 35, 23, 49, 27, 51, 38]
```

We do the same for the sodium and potassium levels:

```
In [6]: sodium = [d['Na'] for d in data]
   ...  potassium = [d['K'] for d in data]
```

These lists can then be passed to Matplotlib's scatter function:

```
In [7]: plt.scatter(sodium, potassium)
   ...  plt.xlabel('sodium')
   ...  plt.ylabel('potassium')
Out[7]: <matplotlib.text.Text at 0x14346584668>
```

This will produce the following plot:

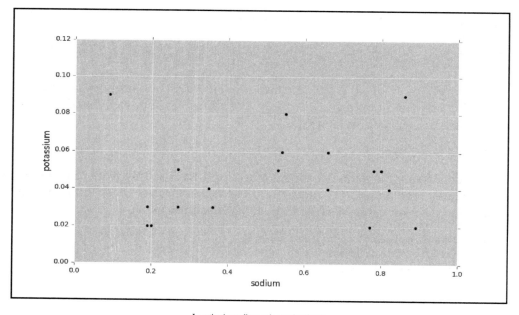

Investigating sodium and potassium levels

However, this plot is not very informative, because all data points have the same color. What we really want is for each data point to be colored according to the drug that was prescribed. For this to work, we need to somehow convert the drug labels `'A'` to `'D'` into numerical values. A nice trick is to use the ASCII value of a character.

If you have done a bit of web development before, you may already know that every character has a corresponding ASCII value, which you can find at http://www.asciitable.com.

In Python, this value is accessible via the function `ord`. For example, the character `'A'` has value 65 (that is, `ord('A') == 65`), `'B'` has 66, `'C'` has 67, and `'D'` has 68. Hence, we can turn the characters `'A'` through `'D'` into integers between 0 and 3 by calling `ord` on them and subtracting 65 from each ASCII value. We do this for every element in the dataset, like we did earlier using a list comprehension:

```
In [8]: target = [ord(t) - 65 for t in target]
   ...: target
Out[8]: [0, 3, 1, 2, 3, 2, 0, 1, 3, 2, 1, 2, 1, 3, 0, 2, 1, 3, 0]
```

We can then pass these integer values to Matplotlib's `scatter` function, which will know to choose different colors for these different color labels (`c=target` in the following code). Let's also increase the size of the dots (`s=100` in the following code) and label our axes, so that we know what we are looking at:

```
In [9]: plt.subplot(221)
   ...: plt.scatter(sodium, potassium, c=target, s=100)
   ...: plt.xlabel('sodium (Na)')
   ...: plt.ylabel('potassium (K)')
   ...: plt.subplot(222)
   ...: plt.scatter(age, potassium, c=target, s=100)
   ...: plt.xlabel('age')
   ...: plt.ylabel('potassium (K)')
   ...: plt.subplot(223)
   ...: plt.scatter(age, sodium, c=target, s=100)
   ...: plt.xlabel('age')
   ...: plt.ylabel('sodium (Na)')
Out[9]: <matplotlib.text.Text at 0x1b36a669e48>
```

The preceding code will produce a figure with four subplots in a 2 x 2 grid, of which the first three subplots show different slices of the dataset, and every data point is colored according to its target label (that is, the prescribed drug):

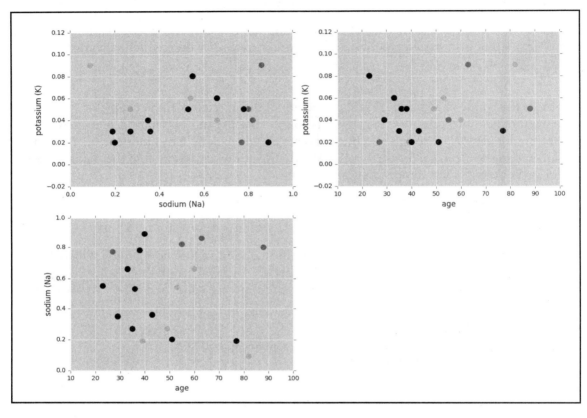

Investing a medical dataset

What do these plots tell us, other than that the dataset is kind of messy? Can you spot any apparent relationship between feature values and target labels?

There are some interesting observations we can make. For example, from the first and third subplot, we can see that the light blue points seem to be clustered around high sodium levels. Similarly, all red points seem to have both low sodium and low potassium levels. The rest is less clear. So let's see how a decision tree would solve the problem.

Preprocessing the data

In order for our data to be understood by the decision tree algorithm, we need to convert all categorical features ('sex', 'BP', and 'cholesterol') into numerical features. What is the best way to do that?

Exactly, using scikit-learn's DictVectorizer. Like we did in the previous chapter, we feed the dataset that we want to convert to the fit_transform method:

```
In [10]: from sklearn.feature_extraction import DictVectorizer
    ...     vec = DictVectorizer(sparse=False)
    ...     data_pre = vec.fit_transform(data)
```

Then, data_pre contains the preprocessed data. If we want to look at the first data point (that is, the first row of data_pre), we match the feature names with the corresponding feature values:

```
In [12]: vec.get_feature_names()
Out[12]: ['BP=high', 'BP=low', 'BP=normal', 'K', 'Na', 'age',
    ...      'cholesterol=high', 'cholesterol=normal',
    ...      'sex=F', 'sex=M']
In [13]: data_pre[0]
Out[13]: array([ 1. , 0. , 0. , 0.06, 0.66, 33. , 1. , 0. ,
                 1. , 0. ])
```

From this, we can see that the three categorical variables, blood pressure ('BP'), cholesterol level ('cholesterol'), and gender ('sex') have been encoded using one-hot coding.

To make sure that our data variables are compatible with OpenCV, we need to convert everything to floating point values:

```
In [14]: import numpy as np
    ...     data_pre = np.array(data_pre, dtype=np.float32)
    ...     target = np.array(target, dtype=np.float32)
```

Then all that's left to do is to split the data into training and tests set, like we did in Chapter 3, *First Steps in Supervised Learning*. Remember that we always want to keep the training and test sets separate. Since we only have 20 data points to work with in this example, we should probably reserve more than 10 percent of the data for testing. A 15-5 split seems appropriate here. We can be explicit and order the `split` function to yield exactly five test samples:

```
In [15]: import sklearn.model_selection as ms
   ...   X_train, X_test, y_train, y_test = 
   ...   ms.train_test_split(data_pre, target, test_size=5,
   ...   random_state=42)
```

Constructing the tree

Building the decision tree with OpenCV works in much the same way as in Chapter 3, *First Steps in Supervised Learning*. Recall that all the machine learning functions reside in OpenCV 3.1's `ml` module. We can create an empty decision tree using the following code:

```
In [16]: import cv2
   ...   dtree = cv2.ml.dtree_create()
```

In order to train the decision tree on the training data, we use the method `train`:

```
In [17]: dtree.train(X_train, cv2.ml.ROW_SAMPLE, y_train)
```

Here we have to specify whether the data samples in `X_train` occupy the rows (using `cv2.ml.ROW_SAMPLE`) or the columns (`cv2.ml.COL_SAMPLE`).

Then we can predict the labels of the new data points with `predict`:

```
In [18]: y_pred = dtree.predict(X_test)
```

If we want to know how good the algorithm is, we can again, use scikit-learn's `accuracy score`:

```
In [19]: from sklearn import metrics
   ...   metrics.accuracy_score(y_test, dtree.predict(X_test))
Out[19]: 0.4
```

This shows us that we only got 40 percent of the test samples right. Since there are only five test samples, this means we effectively got two out of the five samples right. Is the same true for the training set?

```
In [20]: metrics.accuracy_score(y_train, dtree.predict(X_train))
Out[20]: 1.0
```

Not at all! We got all the training samples right. This is typical for decision trees; they tend to learn the training set really well, but then fail to generalize to new data points (such as `X_test`). This is also known as **overfitting**, and we will talk about it in just a second.

But first we have to understand what went wrong. How was the decision tree able to get all the training samples right? How did it do that? And what does the resulting tree actually look like?

Visualizing a trained decision tree

OpenCV's implementation of decision trees is good enough if you are just starting out, and don't care too much about what's going on under the hood. However, in the following sections, we will switch to scikit-learn. Its implementation allows us to customize the algorithm and makes it a lot easier to investigate the inner workings of the tree. Its usage is also much better documented.

In scikit-learn, decision trees can be used for both classification and regression. They reside in the `tree` module:

```
In [21]: from sklearn import tree
```

Similar to OpenCV, we then create an empty decision tree using the `DecisionTreeClassifier` constructor:

```
In [22]: dtc = tree.DecisionTreeClassifier()
```

The tree can then be trained using the `fit` method:

```
In [23]: dtc.fit(X_train, y_train)
Out[23]: DecisionTreeClassifier(class_weight=None, criterion='gini',
            max_depth=None, max_features=None, max_leaf_nodes=None,
            min_impurity_split=1e-07, min_samples_leaf=1,
            min_samples_split=2, min_weight_fraction_leaf=0.0,
            presort=False, random_state=None, splitter='best')
```

Using Decision Trees to Make a Medical Diagnosis

We can then compute the accuracy score on both the training and the test set using the `score` method:

```
In [24]: dtc.score(X_train, y_train)
Out[24]: 1.0
In [25]: dtc.score(X_test, y_test)
Out[25]: 0.40000000000000002
```

Now, here's the cool thing: if you want to know what the tree looks like, you can do so using GraphViz to create a PDF file (or any other supported file type) from the tree structure. For this to work, you need to install GraphViz first, which you can do from the command line using `conda`:

```
$ conda install graphviz
```

Then, back in Python, you can export the tree in GraphViz format to a file `tree.dot` using scikit-learn's `export_graphviz` exporter:

```
In [26]: with open("tree.dot", 'w'):
   ...:     f = tree.export_graphviz(clf, out_file=f)
```

Then, back on the command line, you can use GraphViz to turn `tree.dot` into (for example) a PNG file:

```
$ dot -Tpng tree.dot -o tree.png
```

Alternatively, you can also specify `-Tpdf` or any other supported image format. The result for the preceding tree looks like this:

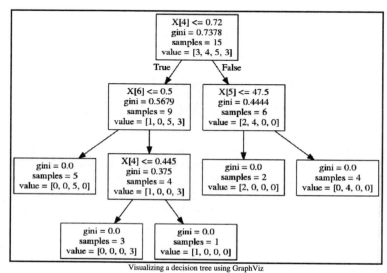

Visualizing a decision tree using GraphViz

What does this all mean? Let's break the figure down step-by-step.

Investigating the inner workings of a decision tree

We established earlier that a decision tree is basically a flow chart that makes a series of decisions about the data. The process starts at the **root node** (which is the node at the very top), where we split the data into two groups, based on some **decision rule**. Then the process is repeated until all remaining samples have the same target label, at which point we have reached a **leaf node**.

In the spam filter example earlier, decisions were made by asking True/False questions. For example, we asked whether an email contained a certain word. If it did, we followed the edge labeled `True` and asked the next question. However, this does not just work for categorical features, but also for numerical features, if we slightly tweak the way we ask a True/False question. The trick is to define a cut-off value for a numerical feature, and then ask questions of the form: *Is feature f larger than value v?* For example, we could have asked whether the word `'prince'` appeared more than five times in an email.

So, what kind of questions were asked in order to arrive at the decision tree plotted earlier?

To get a better understanding, I have annotated the earlier graph with some important components of the decision tree algorithm in the following figure:

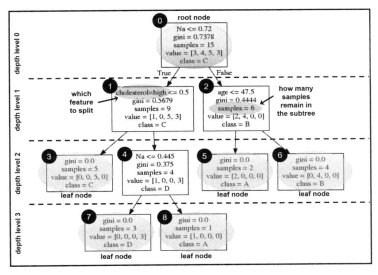

The decision tree graph explained

Starting at the root node (which I have labeled **node 0**), you can see that the first question asked was *whether the sodium concentration was smaller or equal to 0.72*. This resulted in two subgroups:

- All data points where *Na <= 0.72* (**node 1**), which was true for 9 data points
- All data points where *Na > 0.72* (**node 2**), which was true for 6 data points

At node 1, the next question asked was whether the remaining data points did *not* have high cholesterol levels (*cholesterol=high < 0.5*), which was true for 5 data points (**node 3**) and false for 4 data points (**node 4**). At node 3, all 5 remaining data points had the same target label, which was drug C (*class = C*), meaning that there was no more ambiguity to resolve. We call such nodes **pure**. Thus, node 3 became a leaf node.

Back at node 4, the next question asked was *whether sodium levels were lower than 0.445 (Na <= 0.445)*, and the remaining 4 data points were split into **node 7** and **node 8**. At this point, both **node 7** and **node 8** became leaf nodes.

You can see how this algorithm could result in really complex structures if you allowed an unlimited number of questions. Even the tree in the preceding figure can be a bit overwhelming—and this one has only depth 3! Deeper trees are even harder to grasp. In real-world examples, a depth of 10 is actually not uncommon.

The smallest possible decision tree consists of a single node, the root node, asking a single True/False question. We say the tree has **depth 0**. A tree of depth 0 is also called a **decision stump**. It has been used productively in the **adaptive boosting** algorithm (see `Chapter 10`, *Combining Different Algorithms Into an Ensemble*).

Rating the importance of features

What I haven't told you yet is how you pick the features along which to split the data. The preceding root node split the data according to *Na <= 0.72*, but who told the tree to focus on sodium first? Also, where does the number 0.72 come from anyway?

Apparently, some features might be more important than others. In fact, scikit-learn provides a function to rate **feature importance**, which is a number between 0 and 1 for each feature, where 0 means *not used at all in any decisions made* and 1 means *perfectly predicts the target*. The feature importances are normalized such that they all sum to 1:

```
In [27]: dtc.feature_importances_
Out[27]: array([ 0.        , 0.        , 0.        , 0.13554217, 0.29718876,
                0.24096386, 0.        , 0.32630522, 0.        , 0.        ])
```

If we remind ourselves of the feature names, it will become clear which feature seems to be the most important. A plot might be most informative:

```
In [28]: plt.barh(range(10), dtc.feature_importances_, align='center',
    ...:         tick_label=vec.get_feature_names())
```

This will result in the following bar plot:

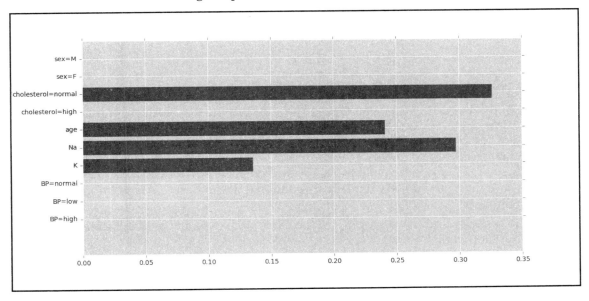

Rating feature importances

Now, it becomes evident that the most telling feature for knowing which drug to administer to patients was actually whether the patient had a normal cholesterol level. Age, sodium levels, and potassium levels were also important. On the other hand, gender and blood pressure did not seem to make any difference at all. However, this does not mean that gender or blood pressure are uninformative. It only means that these features were not picked by the decision tree, likely because another feature would have led to the same splits.

But, hold on. If cholesterol level is so important, why was it not picked as the first feature in the tree (that is, in the root node)? Why would you choose to split on the sodium level first? This is where I need to tell you about that ominous `'gini'` label in the figure earlier.

Feature importances tell us which features are important for classification, but not which class label they are indicative of. For example, we only know that the cholesterol level is important, but we don't know how that led to different drugs being prescribed. In fact, there might not be a simple relationship between features and classes.

Understanding the decision rules

In order to build the perfect tree, you would want to split the tree at the most informative feature, resulting in the purest daughter nodes. However, this simple idea leads to some practical challenges:

- It's not actually clear what most informative means. We need a concrete value, a score function, or a mathematical equation that can describe how informative a feature is.
- In order to find the best split, we have to search over all the possibilities at every decision node.

Fortunately, the decision tree algorithm actually does these two steps for you. Two of the most commonly used criteria that scikit-learn supports are the following:

- `criterion='gini'`: The **Gini impurity** is a measure of misclassification, with the aim of minimizing the probability of misclassification. A perfect split of the data, where each subgroup contains data points of a single target label, would result in a Gini index of 0. We can measure the Gini index of every possible split of the tree, and then choose the one that yields the lowest Gini impurity.

- `criterion='entropy'` (also known as **information gain**): In information theory, **entropy** is a measure of the amount of **uncertainty** associated with a signal or distribution. A perfect split of the data would have 0 entropy. We can measure the entropy of every possible split of the tree, and then choose the one that yields the lowest entropy.

In scikit-learn, you can specify the split criterion in the constructor of the decision tree call. For example, if you want to use entropy, you would type the following:

```
In [29]: dtc_entropy = tree.DecisionTreeClassifier(criterion='entropy')
```

Controlling the complexity of decision trees

If you continue to grow a tree until all leaves are pure, you will typically arrive at a tree that is too complex to interpret. The presence of pure leaves means that the tree is 100 percent correct on the training data, as was the case with our tree shown earlier. As a result, the tree is likely to perform very poorly on the test dataset, as was the case with our tree shown earlier. We say the tree **overfits** the training data.

There are two common ways to avoid overfitting:

- **Pre-pruning**: This is the process of stopping the creation of the tree early
- **Post-pruning** (or just **pruning**): This is the process of first building the tree but then removing or collapsing nodes that contain only little information

There are a number of ways to pre-prune a tree, all of which can be achieved by passing optional arguments to the `DecisionTreeClassifier` constructor:

- Limiting the maximum depth of the tree via the `max_depth` parameter
- Limiting the maximum number of leaf nodes via `max_leaf_nodes`
- Requiring a minimum number of points in a node to keep splitting it via `min_samples_split`

Often pre-pruning is sufficient to control overfitting.

Try it out on our toy dataset! Can you get the score on the test set to improve at all? How does the tree layout change as you start playing with the parameters earlier? If you want to compare answers, check out the Jupyter Notebook on GitHub at: `05.01-Building-Our-First-Decision-Tree.ipynb`.

In more complicated real-world scenarios, pre-pruning is no longer sufficient to control overfitting. In such cases, we want to combine multiple decision trees into what is known as a **random forest**. We will talk about this in `Chapter 10`, *Combining Different Algorithms into an Ensemble*.

Using decision trees to diagnose breast cancer

Now that we have built our first decision tree, it's time to turn our attention to a real dataset: the **Breast Cancer Wisconsin** dataset (`https://archive.ics.uci.edu/ml/datasets/Breast+Cancer+Wisconsin+(Diagnostic)`).

This dataset is a direct result of medical imaging research, and is considered a classic today. The dataset was created from digitized images of healthy (benign) and cancerous (malignant) tissues. Unfortunately, I wasn't able to find any public-domain examples from the original study, but the images look similar to the following one:

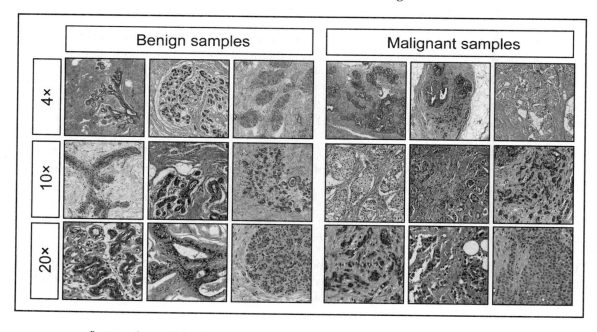

Breast cancer tissue samples from Levenson et al. (2015), PLOS ONE, doi:10.1371/journal.pone.0141357. Released under CC-BY.

The goal of the research was to classify tissue samples into benign and malignant (a binary classification task).

In order to make the classification task feasible, the researchers performed feature extraction on the images, like we did in Chapter 4, *Representing Data and Engineering Features*. They went through a total of 569 images, and extracted 30 different features that described the characteristics of the cell nuclei present in the images, including:

- cell nucleus texture (represented by the standard deviation of the gray scale values)
- cell nucleus size (calculated as the mean of distances from center to points on the perimeter)
- tissue smoothness (local variation in radius lengths)
- tissue compactness

Let's put our newly gained knowledge to good use and build a decision tree to do the classification!

Loading the dataset

The full dataset is part of scikit-learn's example datasets. We can import it using the following commands:

```
In [1]: from sklearn import datasets
   ...     data = datasets.load_breast_cancer()
```

As in the previous examples, all data is contained in a 2D feature matrix `data.data`, where the rows represent data samples and the columns are the feature values:

```
In [2]: data.data.shape
Out[2]: (569, 30)
```

With a look at the provided feature names, we recognize some that we mentioned earlier:

```
In [3]: data.feature_names
Out[3]: array(['mean radius', 'mean texture', 'mean perimeter',
               'mean area', 'mean smoothness', 'mean compactness',
               'mean concavity', 'mean concave points',
               'mean symmetry', 'mean fractal dimension',
               'radius error', 'texture error', 'perimeter error',
               'area error', 'smoothness error',
               'compactness error', 'concavity error',
               'concave points error', 'symmetry error',
               'fractal dimension error', 'worst radius',
```

```
                       'worst texture', 'worst perimeter', 'worst area',
                       'worst smoothness', 'worst compactness',
                       'worst concavity', 'worst concave points',
                       'worst symmetry', 'worst fractal dimension'],
                      dtype='<U23')
```

Since this is a binary classification task, we expect to find exactly two target names:

```
In [4]: data.target_names
Out[4]: array(['malignant', 'benign'], dtype='<U9')
```

Let's reserve some 20 percent of all data samples for testing:

```
In [5]: import sklearn.model_selection as ms
   ...: X_train, X_test, y_train, y_test = 
   ...: ms.train_test_split(data_pre, target, test_size=0.2,
   ...: random_state=42)
```

You could certainly choose a different ratio, but most commonly, people use something like 70-30, 80-20, and 90-10. It all depends a bit on the dataset size, but in the end should not make too much of a difference. Splitting the data 80-20 should result in the following set sizes:

```
In [6]: X_train.shape, X_test.shape
Out[6]: ((455, 30), (114, 30))
```

Building the decision tree

As shown earlier, we can create a decision tree using scikit-learn's tree module. For now, let's not specify any optional arguments:

```
In [5]: from sklearn import tree
   ...: dtc = tree.DecisionTreeClassifier()
```

Do you remember how to train the decision tree?

```
In [6]: dtc.fit(X_train, y_train)
Out[6]: DecisionTreeClassifier(class_weight=None, criterion='gini',
                               max_depth=None, max_features=None,
                               max_leaf_nodes=None,
                               min_impurity_split=1e-07,
                               min_samples_leaf=1,
                               min_samples_split=2,
                               min_weight_fraction_leaf=0.0,
                               presort=False, random_state=None,
                               splitter='best')
```

Since we did not specify any pre-pruning parameters, we would expect this decision tree to grow quite large and result in a perfect score on the training set:

```
In [7]: dtc.score(X_train, y_train)
Out[7]: 1.0
```

However, to our surprise, the test error is not too shabby either:

```
In [8]: dtc.score(X_test, y_test)
Out[8]: 0.94736842105263153
```

And like earlier, we can find out what the tree looks like using `graphviz`:

```
In [9]: with open("tree.dot", 'w') as f:
   ...:     f = tree.export_graphviz(dtc, out_file=f,
   ...:                               feature_names=data.feature_names,
   ...:                               class_names=data.target_names)
```

Indeed, the resulting tree is much more complicated than in our previous example:

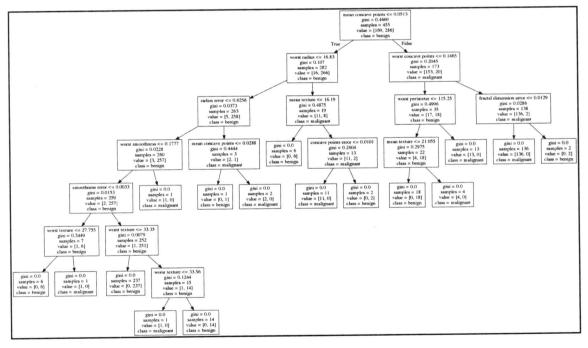

A decision tree for diagnosing breast cancer

At the very top, we find the feature with the name **mean concave points** to be the most informative, resulting in a split that separates the data into a group of 282 potentially benign samples and a group of 173 potentially malignant samples. We also realize that the tree is rather asymmetric, reaching depth 6 on the left but only depth 2 on the right.

The figure of the preceding tree gives us a good baseline performance. Without doing much, we already achieved 94.7 percent accuracy on the test set, thanks to scikit-learn's excellent default parameter values. But would it be possible to get an even higher score on the test set?

To answer this question, we can do some model exploration. For example, we mentioned earlier that the depth of a tree influences its performance. If we want to study this dependency more systematically, we can repeat building the tree for different values of `max_depth`:

```
In [10]: import numpy as np
    ...     max_depths = np.array([1, 2, 3, 5, 7, 9, 11])
```

For each of these values, we want to run the full model cascade from start to finish. We also want to record the train and test scores. We do this in the `for loop`:

```
In [11]: train_score = []
    ...     test_score = []
    ...     for d in max_depths:
    ...         dtc = tree.DecisionTreeClassifier(max_depth=d)
    ...         dtc.fit(X_train, y_train)
    ...         train_score.append(dtc.score(X_train, y_train))
    ...         test_score.append(dtc.score(X_test, y_test))
```

Here, we create a decision tree for every value in `max_depths`, train the tree on the data, and build lists of all the train and test scores. We can plot the scores as a function of the tree depth using Matplotlib:

```
In [12]: import matplotlib.pyplot as plt
    ...     %matplotlib inline
    ...     plt.style.use('ggplot')
In [13]: plt.figure(figsize=(10, 6))
    ...     plt.plot(max_depths, train_score, 'o-', linewidth=3,
    ...              label='train')
    ...     plt.plot(max_depths, test_score, 's-', linewidth=3,
    ...              label='test')
    ...     plt.xlabel('max_depth')
    ...     plt.ylabel('score')
    ...     plt.legend()
Out[13]: <matplotlib.legend.Legend at 0x1c6783d3ef0>
```

This will produce the following plot:

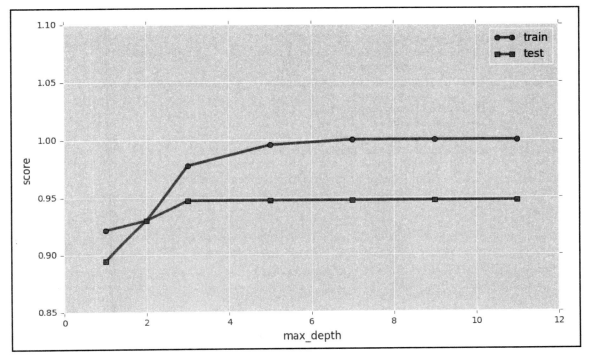

Performance as a function of tree depth

Now it becomes evident how the tree depth influences performance. It seems, the deeper the tree, the better the performance on the training set. Unfortunately, things seem a little more mixed when it comes to the test set performance. Increasing the depth beyond value 3 does not further improve the test score, so we're still stuck at 94.7 percent. Perhaps there is a different pre-pruning setting we could take advantage of that would work better?

Let's do one more. What about the minimum number of samples required to make a node a leaf node?

We repeat the procedure from the previous example:

```
In [14]: train_score = []
    ...  test_score = []
    ...  min_samples = np.array([2, 4, 8, 16, 32])
    ...  for s in min_samples:
    ...      dtc = tree.DecisionTreeClassifier(min_samples_leaf=s)
    ...      dtc.fit(X_train, y_train)
    ...      train_score.append(dtc.score(X_train, y_train))
```

```
...         test_score.append(dtc.score(X_test, y_test))
```

Then we plot the result:

```
In [15]: plt.figure(figsize=(10, 6))
...      plt.plot(min_samples, train_score, 'o-', linewidth=3,
...              label='train')
...      plt.plot(min_samples, test_score, 's-', linewidth=3,
...              label='test')
...      plt.xlabel('max_depth')
...      plt.ylabel('score')
....     plt.legend()
Out[15]: <matplotlib.legend.Legend at 0x1c679914fd0>
```

This leads to a plot that looks quite different from the one before:

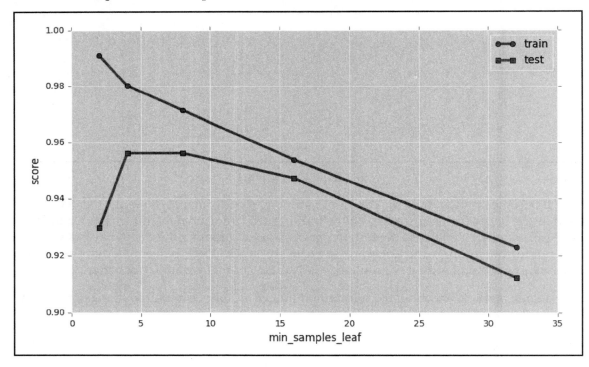

Performance as a function of the minimum number of samples needed to create a leaf node

Clearly, increasing `min_samples_leaf` does not bode well with the training score. But that's not necessarily a bad thing! Because an interesting thing happens in the blue curve: the test score goes through a maximum for values between 4 and 8, leading to the best test score we have found so far-95.6 percent! We just increased our score by 0.9 percent, simply by tweaking the model parameters.

I encourage you to keep tweaking! A lot of great results in machine learning actually come from long-winded hours of trial-and-error model explorations. Before you generate a plot, try thinking about: What you would expect the plot to look like? How should the training score change as you start to restrict the number of leaf nodes (`max_leaf_nodes`)? What about `min_samples_split`? Also, how do things change when you switch from Gini index to information gain?

You will see these kinds of curves quite commonly when trying to tweak model parameters. They tend to have a monotonic relationship on the training score, either steadily improving or worsening it. With test scores, there is usually a sweet spot, a local maximum, after which the test score starts decreasing again.

Using decision trees for regression

Although we have so far focused on using decision trees in classification tasks, you can also use them for regression. But you will need to use scikit-learn again, as OpenCV does not provide this flexibility. We therefore only briefly review its functionality here.

Let's say we wanted to use a decision tree to fit a `sin` wave. To make things interesting, we will also add some noise to the data points using NumPy's random number generator:

```
In [1]: import numpy as np
   ...     rng = np.random.RandomState(42)
```

We then create 100 x values between 0 and 5, and calculate the corresponding `sin` values:

```
In [2]: X = np.sort(5 * rng.rand(100, 1), axis=0)
   ...     y = np.sin(X).ravel()
```

We then add noise to every other data point in y (using `y[::2]`), scaled by 0.5 so we don't introduce too much jitter:

```
In [3]: y[::2] += 0.5 * (0.5 - rng.rand(50))
```

You can then create regression tree like any other tree before. A small difference is that the split criteria `'gini'` and `'entropy'` do not apply to regression tasks. Instead, scikit-learn provides two different split criteria:

- `'mse'` (also known as **variance reduction**): This criterion calculates the **mean squared error** (**MSE**) between ground truth and prediction, and splits the node that leads to the smallest MSE
- `'mae'`: This criterion calculates the **mean absolute error** (**MAE**) between ground truth and prediction, and splits the node that leads to the smallest MAE

Using the `'mse'` criterion, we will build two trees; one with a depth of 2 and one with a depth of 5:

```
In [4]: from sklearn import tree
In [5]: regr1 = tree.DecisionTreeRegressor(max_depth=2,
...         random_state=42)
...     regr1.fit(X, y)
Out[5]: DecisionTreeRegressor(criterion='mse', max_depth=2,
                    max_features=None, max_leaf_nodes=None,
                    min_impurity_split=1e-07,
                    min_samples_leaf=1, min_samples_split=2,
                    min_weight_fraction_leaf=0.0,
                    presort=False, random_state=42,
                    splitter='best')
In [6]: regr2 = tree.DecisionTreeRegressor(max_depth=5,
...         random_state=42)
...     regr2.fit(X, y)
Out[6]: DecisionTreeRegressor(criterion='mse', max_depth=5,
                    max_features=None, max_leaf_nodes=None,
                    min_impurity_split=1e-07,
                    min_samples_leaf=1, min_samples_split=2,
                    min_weight_fraction_leaf=0.0,
                    presort=False, random_state=42,
                    splitter='best')
```

We can then use the decision tree like a linear regressor from Chapter 3, *First Steps in Supervised Learning*. For this, we create a test set with *x* values densely sampled in the whole range from 0 through 5:

```
In [7]: X_test = np.arange(0.0, 5.0, 0.01)[:, np.newaxis]
```

The predicted *y* values can then be obtained with the `predict` method:

```
In [8]: y_1 = regr1.predict(X_test)
...     y_2 = regr2.predict(X_test)
```

If we plot all of these together, we can see how the decision trees differ:

```
In [9]: import matplotlib.pyplot as plt
   ...  %matplotlib inline
   ...  plt.style.use('ggplot')

   ...  plt.scatter(X, y, c='k', s=50, label='data')
   ...  plt.plot(X_test, y_1, label="max_depth=2", linewidth=5)
   ...  plt.plot(X_test, y_2, label="max_depth=5", linewidth=3)
   ...  plt.xlabel("data")
   ...  plt.ylabel("target")
   ...  plt.legend()
Out[9]: <matplotlib.legend.Legend at 0x12d2ee345f8>
```

This will produce the following plot:

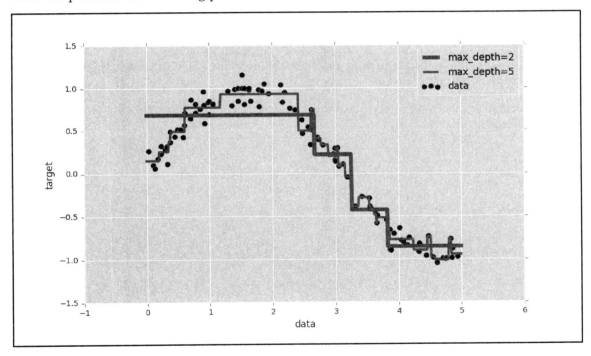

Using decision trees for regression

Here, the thick line represents the regression tree with depth 2. You can see how the tree tries to approximate the data using these crude steps. The thinner line belongs to the regression tree with depth 5; the added depth has allowed the tree to make much finer grained decisions. Therefore, this tree is able to approximate the data even better. However, because of this added power, the tree is also more susceptible to fitting noisy values, as can be seen especially on the right-hand side of the plot.

Summary

If you ask me, this chapter went by really fast. We learned all about decision trees, and how to apply them to both classification and regression tasks. We talked a bit about overfitting, and ways to avoid this phenomenon by tweaking pre-pruning and post-pruning settings. We also learned about how to rate the quality of a node split using metrics such as the Gini impurity and information gain. Finally, we applied decision trees to medical data in order to detect cancerous tissues. We will come back to decision trees towards the end of the book, when we will combine multiple trees into what is known as a random forest. But for now, let's move on to a new topic.

In the next chapter, we will introduce another staple of the machine learning world: support vector machines.

6
Detecting Pedestrians with Support Vector Machines

In the previous chapter, we talked about how to use decision trees for classification and regression. In this chapter, we want to direct our attention to another well established supervised learner in the machine learning world: **support vector machines** (**SVMs**).

Soon after their introduction in the early 1990, SVMs quickly became popular in the machine learning community, largely because of their success in early **handwritten digit classification**. They remain relevant to this day, especially in application domains, such as computer vision.

The goal of this chapter is to apply SVMs to a popular problem in computer vision: **pedestrian detection**. In contrast to a recognition task (where we name the category of an object), the goal of a detection task is to say whether a particular object (or in our case, a pedestrian) is present in an image or not.

You might already know that OpenCV can do this in two to three lines of code. But, we won't learn anything if we do it that way. So instead, we'll build the whole pipeline from scratch! We will obtain a real-world dataset, perform feature extraction using the **histogram of oriented gradients** (**HOG**), and apply an SVM to it.

Along the way, we want to address the following questions:

- How do we implement SVMs in OpenCV with Python, and how do they work?
- What's the deal with decision boundaries, and why are SVMs also called maximum-margin classifiers?
- What is the kernel trick, and how do we put it into practice?

Excited? Then let's go!

Understanding linear support vector machines

In order to understand how SVMs work, we have to think about **decision boundaries**. When we used linear classifiers or decision trees in earlier chapters, our goal was always to minimize the classification error. We did this by accuracy or mean squared error. An SVM tries to achieve low classification errors too, but it does so only implicitly. An SVM's explicit objective is to maximize the **margins** between data points of one class versus the other. This is the reason SVMs are sometimes also called **maximum-margin classifiers**.

Learning optimal decision boundaries

Let's look at a simple example. Consider some training samples with only two features (x and y values) and a corresponding target label (positive (+) or negative (-)). Since the labels are categorical, we know that this is a classification task. Moreover, because we only have two distinct classes (+ and -), it's a **binary classification task**.

In a binary classification task, a decision boundary is a line that **partitions** the training set into two subsets, one for each class. An **optimal decision boundary** partitions the data such that all data samples from one class (say, +) are to the left of the decision boundary, and all other data samples (say, -) are to the right of it.

An SVM updates its choice of a decision boundary throughout the training procedure. For example, at the beginning of training, the classifier has seen only a few data points, and it tries to draw in the decision boundary that best separates the two classes. As training progresses, the classifier sees more and more data samples, and so it keeps updating the decision boundary at each step. This process is illustrated in the following figure:

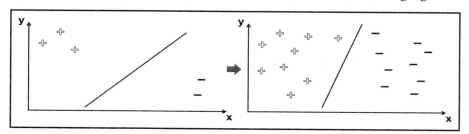

The decision boundary is constantly updated during the training phase

As training progresses, the classifier gets to see more and more data samples, and thus gets an increasingly better idea of where the optimal decision boundary should lie. In this scenario, a **misclassification error** would happen if the decision boundary were drawn in such a way that a — sample were to the left of it, or a + were to the right of it.

We know from the previous chapter that after training, the classification model is no longer tampered with. This means that the classifier will have to predict the target label of new data points using the decision boundary it obtained during training.

In other words, during testing, we want to know in the following figure, which class the ? sign should have, based on the decision boundary we learned during the training phase:

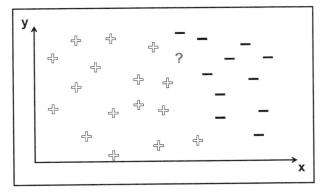

With the decision boundary we learned during training, can we predict the target label of new data points?

You can see why this is usually a tricky problem. If the location of the question mark was more to the left, we would have been certain that the corresponding target label is +. However, in this case, there are several ways to draw the decision boundary, such that all the + samples are to the left of it and all - samples are to the right of it, as illustrated in this figure:

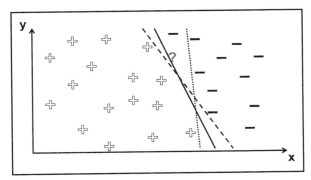

Multiple decision boundaries can do the job, but which one is the best?

Which decision boundary would you choose?

All three decision boundaries get the job done; they perfectly split the training data into + and - subsets without making any misclassification errors. However, depending on the decision boundary we choose, the ? would come to lie either on the left of it (dotted and solid lines), in which case it would be assigned to the + class, or to the right of it (dashed line), in which case it would be assigned to the - class.

This is what SVMs are really good at. An SVM would most likely pick the solid line, because this is the decision boundary that maximizes the margin between the data points from the + and - classes. This is illustrated in the following figure:

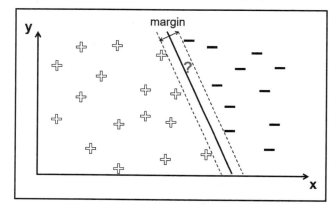

Example of a decision boundary an SVM would learn

 It turns out that in order to find the maximal margin, it is only necessary to consider the data points that lie on the class margins. These data points are also called the support vectors. That's where SVMs got their name from.

Implementing our first support vector machine

But enough with the theory. Let's do some coding!

However, it might be a good idea to pace ourselves. For our very first SVM, we should probably focus on a simple dataset, perhaps a binary classification task.

A cool trick about scikit-learn's dataset module that I haven't told you about is that you can generate random datasets of controlled size and complexity. A few notable ones are:

- `datasets.make_classification([n_samples, ...])`: This function generates a random n-class classification problem, where we can specify the number of samples, the number of features, and the number of target labels
- `datasets.make_regression([n_samples, ...])`: This function generates a random regression problem
- `datasets.make_blobs([n_samples, n_features, ...])`: This function generates a number of Gaussian blobs we can use for clustering

This means that we can use `make_classification` to build a custom dataset for a binary classification task.

Generating the dataset

As we can now recite in our sleep, a binary classification problem has exactly two distinct target labels (n_classes=2). For the sake of simplicity, let's limit ourselves to only two feature values (n_features=2; for example, an *x* and a *y* value). Let's say we want to create 100 data samples:

```
In [1]: from sklearn import datasets
   ...   X, y = datasets.make_classification(n_samples=100, n_features=2,
   ...                                        n_redundant=0, n_classes=2,
   ...                                        random_state=7816)
```

We expect X to have 100 rows (data samples) and two columns (features), whereas the vector y should have a single column that contains all the target labels:

```
In [2]: X.shape, y.shape
Out[2]: ((100, 2), (100,))
```

Visualizing the dataset

We can plot these data points in a scatter plot using Matplotlib. Here, the idea is to plot the *x* values (found in the first column of X, X[:, 0]) against the *y* values (found in the second column of X, X[:, 1]). A neat trick is to pass the target labels as color values (c=y):

```
In [3]: import matplotlib.pyplot as plt
   ...   %matplotlib inline
   ...   plt.scatter(X[:, 0], X[:, 1], c=y, s=100)
   ...   plt.xlabel('x values')
```

```
...         plt.ylabel('y values')
Out[3]: <matplotlib.text.Text at 0x24f7ffb00f0>
```

This will produce the following figure:

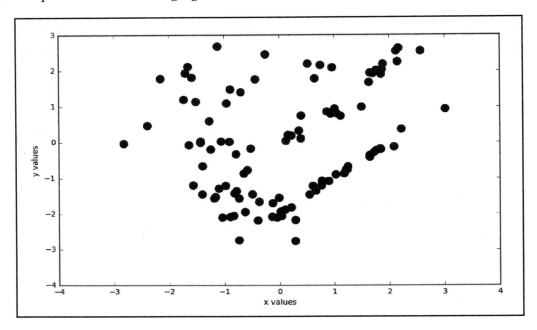

Randomly generated data for a binary classification problem

You can see that, for the most part, data points of the two classes are clearly separated. However, there are a few regions (particularly near the left and bottom of the plot) where the data points of both classes intermingle. These will be hard to classify correctly, as we will see in just a second.

Preprocessing the dataset

The next step is to split the data points into training and test sets, as we have done before. But, before we do that, we have to prepare the data for OpenCV:

- All feature values in X must be 32-bit floating point numbers
- Target labels must be either -1 or +1

We can achieve this with the following code:

```
In [4]: import numpy as np
   ...  X = X.astype(np.float32)
   ...  y = y * 2 - 1
```

Now we can pass the data to scikit-learn's `train_test_split` function, like we did in the earlier chapters:

```
In [5]: from sklearn import model_selection as ms
   ...  X_train, X_test, y_train, y_test = ms.train_test_split(
   ...      X, y, test_size=0.2, random_state=42
   ...  )
```

Here I chose to reserve 20 percent of all data points for the test set, but you can adjust this number according to your liking.

Building the support vector machine

In OpenCV, SVMs are built, trained, and scored the same exact way as every other learning algorithm we have encountered so far, using the following four steps:

1. Call the `create` method to construct a new SVM:

   ```
   In [6]: import cv2
      ...  svm = cv2.ml.SVM_create()
   ```

   As shown in the following command, there are different *modes* in which we can operate an SVM. For now, all we care about is the case we discussed in the previous example: an SVM that tries to partition the data with a straight line. This can be specified with the `setKernel` method:

   ```
   In [7]: svm.setKernel(cv2.ml.SVM_LINEAR)
   ```

2. Call the classifier's `train` method to find the optimal decision boundary:

   ```
   In [8]: svm.train(X_train, cv2.ml.ROW_SAMPLE, y_train)
   Out[8]: True
   ```

3. Call the classifier's `predict` method to predict the target labels of all data samples in the test set:

   ```
   In [9]: _, y_pred = svm.predict(X_test)
   ```

4. Use scikit-learn's `metrics` module to score the classifier:

```
In [10]: from sklearn import metrics
   ...:  metrics.accuracy_score(y_test, y_pred)
Out[10]: 0.80000000000000004
```

Congratulations, we got 80 percent correctly classified test samples!

Of course, so far we have no idea what happened under the hood. For all we know, we might as well have gotten these commands off a web search and typed them into the terminal, without really knowing what we're doing. But this is not who we want to be. Getting a system to work is one thing and understanding it is another. Let us get to that!

Visualizing the decision boundary

What was true in trying to understand our data is true for trying to understand our classifier: visualization is the first step in understanding a system. We know the SVM somehow came up with a decision boundary that allowed us to correctly classify 80 percent of the test samples. But how can we find out what that decision boundary actually looks like?

For this, we will borrow a trick from the guys behind scikit-learn. The idea is to generate a fine grid of x and y coordinates and run that through the SVM's `predict` method. This will allow us to know, for every (x, y) point, what target label the classifier would have predicted.

We will do this in a dedicated function, which we call `plot_decision_boundary`. The function takes as inputs an SVM object, the feature values of the test set, and the target labels of the test set:

```
In [11]: def plot_decision_boundary(svm, X_test, y_test):
```

To generate the grid (also called **mesh grid**), we first have to figure out how much space the data samples in the test set take up in the x-y plane. To find the leftmost point on the plane, we look for the smallest x value in `X_test`, and to find the rightmost plane, we look for the largest x value in `X_test`. We don't want any data points to fall on the border, so we add some margin of + 1 or - 1:

```
   ...:         x_min, x_max = X_test[:, 0].min() - 1, X_test[:, 0].max() + 1
```

We do the same for y:

```
...         y_min, y_max = X_test[:, 1].min() - 1, X_test[:, 1].max() + 1
```

From these boundary values, we can then create a fine mesh grid (with sampling step h):

```
...         h = 0.02   # step size in mesh
...         xx, yy = np.meshgrid(np.arange(x_min, x_max, h),
...                              np.arange(y_min, y_max, h))
...         X_hypo = np.c_[xx.ravel().astype(np.float32),
...                        yy.ravel().astype(np.float32)]
```

Here we make use of NumPy's `arange(start, stop, step)` function that creates linearly spaced values between `start` and `stop`, all `step` apart.

We want to treat each of these (xx, yy) coordinates as hypothetical data points. So we stack them column-wise into a N x 2 matrix:

```
...         X_hypo = np.c_[xx.ravel().astype(np.float32),
...                        yy.ravel().astype(np.float32)]
...         _, zz = svm.predict(X_hypo)
```

Don't forget to convert the values to 32-bit floating point numbers again! Otherwise, OpenCV will complain.

Now we can pass the `X_hypo` matrix to the `predict` method:

```
...         _, zz = svm.predict(X_hypo)
```

The resulting target labels `zz` will be used to create a colormap of the feature landscape:

```
...         zz = zz.reshape(xx.shape)
...         plt.contourf(xx, yy, zz, cmap=plt.cm.coolwarm, alpha=0.8)
```

This creates a contour plot, on top of which we will plot the individual data points colored by their true target labels:

```
...         plt.scatter(X_test[:, 0], X_test[:, 1], c=y_test, s=200)
```

The function can be called with the following code:

```
In [12]: plot_decision_boundary(svm, X_test, y_test)
```

The result looks like this:

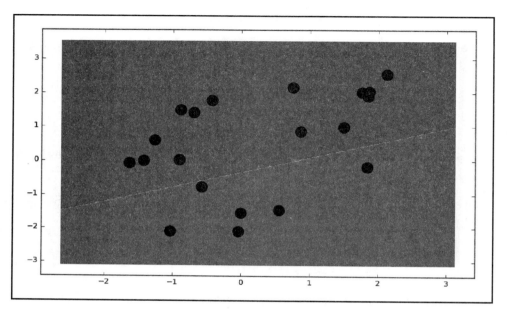

Visualizing a linear decision boundary

Now we get a better sense of what is going on!

The SVM found a straight line (a linear decision boundary) that best separates the blue and the red data samples. It didn't get all the data points right, as there are three blue dots in the red zone and one red dot in the blue zone.

However, we can convince ourselves that this is the best straight line we could have chosen by wiggling the line around in our heads:

- If we rotate the line to make it more horizontal, we might end up misclassifying the rightmost blue dot—the one at coordinates (-2, 0)—which could come to lie in the red region right over the the horizontal line.
- If we keep rotating the line in an effort to make the three blue dots on the left fall in the blue zone, we will unavoidably also put the one red dot that is currently over the decision boundary, the one at coordinates (-1, 0), into the blue zone
- If we make more drastic changes to the decision boundary, and make the straight line almost vertical, in an effort to correctly classify the three blue dots on the left, we end up putting the blue dots in the lower right into the red zone

Thus, no matter how we wiggle and rotate the line, if we end up correctly classifying some currently misclassified points, we also end up misclassifying some other points that are currently correctly classified. It's a vicious cycle!

So what can we do to improve our classification performance?

One solution is to move away from straight lines and onto more complicated decision boundaries.

Dealing with nonlinear decision boundaries

What if the data cannot be optimally partitioned using a linear decision boundary? In such a case, we say the data is not linearly separable.

The basic idea to deal with data that is not linearly separable is to create **nonlinear combinations** of the original features. This is the same as saying we want to project our data to a higher-dimensional space (for example, from 2D to 3D) in which the data suddenly becomes linearly separable. This concept is illustrated in the following figure:

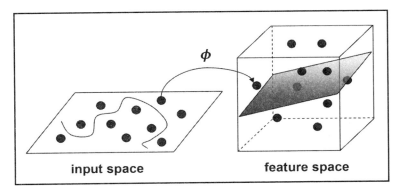

Finding linear hyperplanes in higher-dimensional spaces

If data in its original input space (left) cannot be linearly separated, we can apply a mapping function $\phi(.)$ that projects the data from 2D into a 3D plane. In this higher-dimensional space, we may find that there is now a linear decision boundary (which, in 3D, is a **plane**) that can separate the data.

A linear decision boundary in an n-dimensional space is called a **hyperplane**. For example, a decision boundary in 6D feature space is a 5D hyperplane; in 3D feature space, it's a regular 2D plane; and in 2D space, it's a straight line.

However, one problem with this mapping approach is that it is impractical in large dimensions, because it adds a lot of extra terms to do the mathematical projections between the dimensions. This is where the so-called **kernel trick** comes in to play.

Understanding the kernel trick

Granted, we won't have time to develop all the mathematics needed to truly understand the kernel trick. A more realistic section title would have been *Acknowledging that something called the kernel trick exists and accepting that it works*, but that would have been a bit wordy.

Here's the kernel trick in a nutshell.

In order to figure out the slope and orientation of the decision hyperplane in the high-dimensional space, we have to multiply all the feature values with appropriate weight values, and sum them all up. The more dimensions our feature space has, the more work we have to do.

However, mathematicians smarter than us have long realized that an SVM has no need to explicitly work in the higher-dimensional space, neither during training nor testing! As it turns out, the optimization problem only uses the training examples to compute pair-wise dot products between two features. This got mathematicians excited, because there is a **trick** to compute such a product without explicitly having to transform the features into the higher-dimensional space. The type of functions that can do this clever computation are called **kernel functions**.

Hence, the kernel trick!

One such class of kernel functions is called **radial basis functions** (**RBFs**). The values of RBFs depend only on the distance from a reference point. An example of an RBF as a function of a radius r would be $f(r) = 1/r$, or $f(r) = 1/r^2$. A more common example is the **Gaussian function** (also known as the **bell curve**).

RBFs can be used to produce nonlinear decision boundaries that group data points into **blobs** or **hotspots**, such as illustrated in the following figure:

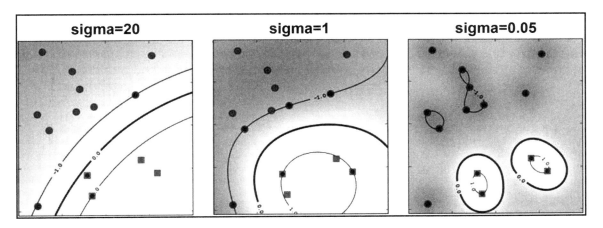

Example Gaussian kernels with different standard deviations (sigma)

By adjusting the standard deviation of the Gaussian function (that is, the fall-off with distance from the center), it is possible to create a great number of intricate decision boundaries, especially in higher dimensions.

Knowing our kernels

OpenCV provides a whole range of different SVM kernels with which we can experiment. Some of the most commonly used ones include:

- `cv2.ml.SVM_LINEAR`: This is the kernel we used previously. It provides a linear decision boundary in the original feature space (the *x* and *y* values).
- `cv2.ml.SVM_POLY`: This kernel provides a decision boundary that is a polynomial function in the original feature space. In order to use this kernel, we also have to specify a coefficient via `svm.setCoef0` (usually set to 0) and the degree of the polynomial via `svm.setDegree`.
- `cv2.ml.SVM_RBF`: This kernel implements the kind of Gaussian function we discussed earlier.
- `cv2.ml.SVM_SIGMOID`: This kernel implements a sigmoid function, similar to the one we encountered when talking about logistic regression in Chapter 3, *First Steps in Supervised Learning*.
- `cv2.ml.SVM_INTER`: This kernel is a new addition to OpenCV 3. It separates classes based on the similarity of their histograms.

Implementing nonlinear support vector machines

In order to test some of the SVM kernels we just talked about, we will return to our code sample mentioned earlier. We want to repeat the process of building and training the SVM on the dataset generated earlier, but this time we want to use a whole range of different kernels:

```
In [13]: kernels = [cv2.ml.SVM_LINEAR, cv2.ml.SVM_INTER,
    ...              cv2.ml.SVM_SIGMOID, cv2.ml.SVM_RBF]
```

Do you remember what all of these stand for?

Setting a different SVM kernel is relatively simple. We take an entry from the `kernels` list and pass it to the `setKernels` method of the SVM class. That's all.

The laziest way to repeat things is to use a `for` loop:

```
In [14]: for idx, kernel in enumerate(kernels):
```

Then the steps are as follows:

1. Create the SVM and set the `kernel` method:

   ```
   ...        svm = cv2.ml.SVM_create()
   ...        svm.setKernel(kernel)
   ```

2. Train the classifier:

   ```
   ...        svm.train(X_train, cv2.ml.ROW_SAMPLE, y_train)
   ```

3. Score the model using scikit-learn's metrics module imported previously:

   ```
   ...        _, y_pred = svm.predict(X_test)
   ...        accuracy = metrics.accuracy_score(y_test, y_pred)
   ```

4. Plot the decision boundary in a 2 x 2 subplot (remember that subplots in Matplotlib are 1-indexed to mimic Matlab, so we have to call `idx + 1`):

   ```
   ...        plt.subplot(2, 2, idx + 1)
   ...        plot_decision_boundary(svm, X_test, y_test)
   ...        plt.title('accuracy = %.2f' % accuracy)
   ```

The result looks like this:

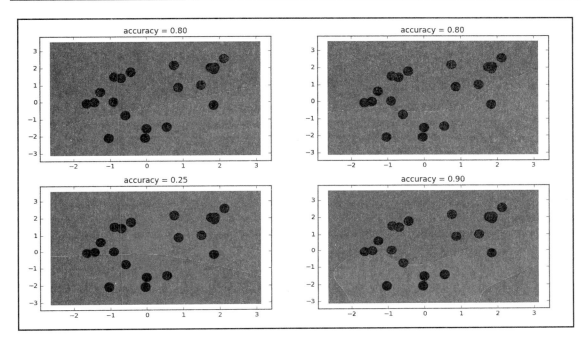

Comparing decision boundaries for different SVM kernels

Let's break the figure down step-by-step:

First, we find that the linear kernel (top-left panel) still looks like one in an earlier plot. We now realize that it's also the only version of SVM that produces a straight line as a decision boundary (although cv2.ml.SVM_C produces almost identical results to cv2.ml.SVM_LINEAR).

The histogram intersection kernel (top-right panel) allows for a more complex decision boundary. However, this did not improve our generalization performance (accuracy is still at 80 percent).

Although the sigmoid kernel (bottom-left panel) allows for a nonlinear decision boundary, it made a really poor choice, leading to only 25% accuracy.

On the other hand, the Gaussian kernel (bottom-right panel) was able to improve our performance to 90% accuracy. It did so by having the decision boundary wrap around the lowest red dot and reaching up to put the two leftmost blue dots into the blue zone. It still makes two mistakes, but it definitely draws the best decision boundary we have seen to date! Also, note how the RBF kernel is the only kernel that cares to narrow down the blue zone in the lower two corners.

Detecting pedestrians in the wild

We briefly talked about the difference between detection and recognition. While recognition is concerned with classifying objects (for example, as pedestrians, cars, bicycles, and so on), detection is basically answering the question: is there a pedestrian present in this image?

The basic idea behind most detection algorithms is to split up an image into many small patches, and then classify each image patch as either containing a pedestrian or not. This is exactly what we are going to do in this section. In order to arrive at our own pedestrian detection algorithm, we need to perform the following steps:

1. Build a database of images containing pedestrians. These will be our positive data samples.
2. Build a database of images not containing pedestrians. These will be our negative data samples.
3. Train an SVM on the dataset.
4. Apply the SVM to every possible patch of a test image in order to decide whether the overall image contains a pedestrian.

Obtaining the dataset

For the purpose of this section, we will work with the MIT People dataset, which we are free to use for non-commercial purposes. So make sure not to use this in your groundbreaking autonomous start-up company before obtaining a corresponding software license.

The dataset can be obtained from `http://cbcl.mit.edu/software-datasets/Pedestrian Data.html`. There you should find a **DOWNLOAD** button that leads you to a file called `http://cbcl.mit.edu/projects/cbcl/software-datasets/pedestrians128x64.tar.gz`.

However, if you followed our installation instructions from earlier and checked out the code on GitHub, you already have the dataset and are ready to go! The file can be found at `notebooks/data/pedestrians128x64.tar.gz`.

Since we are supposed to run this code from a Jupyter Notebook in the `notebooks/` directory, the relative path to the data directory is simply `data/`:

```
In [1]: datadir = "data"
   ...: dataset = "pedestrians128x64"
   ...: datafile = "%s/%s.tar.gz" % (datadir, dataset)
```

The first thing to do is to unzip the file. We will extract all files into their own subdirectories in `data/pedestrians128x64/`:

```
In [2]: extractdir = "%s/%s" % (datadir, dataset)
```

We can do this either by hand (outside of Python) or with the following function:

```
In [3]: def extract_tar(filename):
...         try:
...             import tarfile
...         except ImportError:
...             raise ImportError("You do not have tarfile installed. "
...                               "Try unzipping the file outside of "
...                               "Python.")
...         tar = tarfile.open(datafile)
...         tar.extractall(path=extractdir)
...         tar.close()
...         print("%s sucessfully extracted to %s", % (datafile,
...                                                    extractdir))
```

Then we can call the function like this:

```
In [4]: extract_tar(datafile)
Out[4]: data/pedestrians128x64.tar.gz successfully extracted to
        data/pedestrians128x64
```

If you haven't seen the `try...except` statement before, it is a great way to deal with exceptions as they occur. For example, trying to import a module called `tarfile` could result in an `ImportError` if the module does not exist. We capture this exception with `except ImportError` and display a custom error message. You can find more information about handling exceptions in the official Python docs at https://docs.python.org/3/tutorial/errors.html#handling-exceptions.

The dataset comes with a total of 924 color images of pedestrians, each scaled to 64 x 128 pixels and aligned so that the person's body is in the center of the image. Scaling and aligning all data samples is an important step of the process, and we are glad that we don't have to do it ourselves.

These images were taken in Boston and Cambridge in a variety of seasons and under several different lighting conditions. We can visualize a few example images by reading the image with OpenCV and passing an RGB version of the image to Matplotlib:

```
In [5]: import cv2
...     import matplotlib.pyplot as plt
...     %matplotlib inline
```

We will do this in a loop for images 100-104 in the dataset, so we get an idea of the different lighting conditions:

```
In [6]: for i in range(5):
   ...      filename = "%s/per0010%d.ppm" % (extractdir, i)
   ...      img = cv2.imread(filename)
   ...      plt.subplot(1, 5, i + 1)
   ...      plt.imshow(cv2.cvtColor(img, cv2.COLOR_BGR2RGB))
   ...      plt.axis('off')
```

The output looks as follows:

Examples of images of people in the database. The examples vary in color, texture, view point (either frontal or rear), and background.

We can look at additional pictures, but it's pretty clear that there is no straightforward way to describe pictures of people as easily as we did in the previous section for + and - data points. Part of the problem is thus finding a good way to represent these images. Does this ring a bell? It should! We're talking about feature engineering.

Taking a glimpse at the histogram of oriented gradients (HOG)

The HOG might just provide the help we're looking for in order to get this project done. HOG is a feature descriptor for images, much like the ones we discussed in Chapter 4, *Representing Data and Engineering Features*. It has been successfully applied to many different tasks in computer vision, but seems to work especially well for classifying people.

The essential idea behind HOG features is that the local shapes and appearance of objects within an image can be described by the distribution of edge directions. The image is divided into small connected regions, within which a histogram of gradient directions (or **edge directions**) is compiled. Then, the descriptor is assembled by concatenating the different histograms. An example is shown in the following image:

HOG descriptor applied to an example image

The picture on the right-hand side shows which edge orientations dominate in different subregions of the image. You can see why this descriptor would be particularly well suited to data that is rich in texture. For improved performance, the local histograms can also be contrast-normalized, which results in better invariance to changes in illumination and shadowing.

The HOG descriptor is fairly accessible in OpenCV by means of cv2.HOGDescriptor, which takes a bunch of input arguments, such as the detection window size (minimum size of the object to be detected, 48 x 96), the block size (how large each box is, 16 x 16), the cell size (8 x 8), and the cell stride (how many pixels to move from one cell to the next, 8 x 8). For each of these cells, the HOG descriptor then calculates a histogram of oriented gradients using nine bins:

```
In [7]: win_size = (48, 96)
   ...: block_size = (16, 16)
   ...: block_stride = (8, 8)
```

```
...         cell_size = (8, 8)
...         num_bins = 9
...         hog = cv2.HOGDescriptor(win_size, block_size, block_stride,
...                                 cell_size, num_bins)
```

Although this function call looks fairly complicated, these are actually the only values for which the HOG descriptor is implemented. The argument that matters the most is the window size (`win_size`).

All that's left to do is call `hog.compute` on our data samples. For this, we build a dataset of positive samples (`X_pos`) by randomly picking pedestrian images from our data directory. In the following code snippet, we randomly select 400 pictures from the over 900 available, and apply the HOG descriptor to them:

```
In [8]: import numpy as np
   ...: import random
   ...: random.seed(42)
   ...: X_pos = []
   ...: for i in random.sample(range(900), 400):
   ...:     filename = "%s/per%05d.ppm" % (extractdir, i)
   ...:     img = cv2.imread(filename)
   ...:     if img is None:
   ...:         print('Could not find image %s' % filename)
   ...:         continue
   ...:     X_pos.append(hog.compute(img, (64, 64)))
```

We should also remember that OpenCV wants the feature matrix to contain 32-bit floating point numbers, and the target labels to be 32-bit integers. We don't mind, since converting to NumPy arrays will allow us to easily investigate the sizes of the matrices we created:

```
In [9]: X_pos = np.array(X_pos, dtype=np.float32)
   ...: y_pos = np.ones(X_pos.shape[0], dtype=np.int32)
   ...: X_pos.shape, y_pos.shape
Out[9]: ((399, 1980, 1), (399,))
```

It looks like we picked a total of 399 training samples, each of which have 1,980 feature values (which are the HOG feature values).

Generating negatives

The real challenge, however, is to come up with the perfect example of a non-pedestrian. After all, it's easy to think of example images of pedestrians. But what is the opposite of a pedestrian?

This is actually a common problem when trying to solve new machine learning problems. Both research labs and companies spend a lot of time creating and annotating new datasets that fit their specific purpose.

If you're stumped, let me give you a hint on how to approach this. A good first approximation to finding the opposite of a pedestrian is to assemble a dataset of images that look like the images of the positive class but do not contain pedestrians. These images could contain anything like cars, bicycles, streets, houses, and maybe even forests, lakes, or mountains.

A good place to start is the Urban and Natural Scene dataset by the Computational Visual Cognition Lab at MIT. The complete dataset can be obtained from http://cvcl.mit.edu/database.htm, but don't bother. I have already assembled a good amount of images from categories such as open country, inner cities, mountains, and forests. You can find them in the data/pedestrians_neg directory:

```
In [10]: negdir = "%s/pedestrians_neg" % datadir
```

All the images are in color, in .jpeg format, and are 256 x 256 pixels. However, in order to use them as samples from a negative class that go together with our images of pedestrians earlier, we need to make sure that all images have the same pixel size. Moreover, the things depicted in the images should roughly be at the same scale. Thus, we want to loop through all the images in the directory (via os.listdir) and cut out a 64 x 128 **region of interest** (**ROI**):

```
In [11]: import os
    ...: hroi = 128
    ...: wroi = 64
    ...: X_neg = []
    ...: for negfile in os.listdir(negdir):
    ...:     filename = '%s/%s' % (negdir, negfile)
```

To bring the images to roughly the same scale as the pedestrian images, we resize them:

```
    ...:     img = cv2.imread(filename)
    ...:     img = cv2.resize(img, (512, 512))
```

Then we cut out a 64 x 128 pixel ROI by randomly choosing the top-left corner coordinate (rand_x, rand_y). We do this an arbitrary number of five times to bolster up our database of negative samples:

```
    ...:     for j in range(5):
    ...:         rand_y = random.randint(0, img.shape[0] - hroi)
    ...:         rand_x = random.randint(0, img.shape[1] - wroi)
    ...:         roi = img[rand_y:rand_y + hroi, rand_x:rand_x + wroi, :]
```

```
            X_neg.append(hog.compute(roi, (64, 64)))
```

Some examples from this procedure are shown in the following image:

Example negatives extracted from natural scenes

What did we almost forget? Exactly, we forgot to make sure that all feature values are 32-bit floating point numbers. Also, the target label of these images should be -1, corresponding to the negative class:

```
In [12]: X_neg = np.array(X_neg, dtype=np.float32)
    ...: y_neg = -np.ones(X_neg.shape[0], dtype=np.int32)
    ...: X_neg.shape, y_neg.shape
Out[12]: ((250, 1980, 1), (250,))
```

Then we can concatenate all positive (X_pos) and negative samples (X_neg) into a single dataset X, which we split using the all too familiar `train_test_split` function from scikit-learn:

```
In [13]: X = np.concatenate((X_pos, X_neg))
    ...: y = np.concatenate((y_pos, y_neg))
In [14]: from sklearn import model_selection as ms
    ...: X_train, X_test, y_train, y_test = ms.train_test_split(
    ...:     X, y, test_size=0.2, random_state=42
    ...: )
```

> A common and painful mistake is to accidentally include a negative sample that does indeed contain a pedestrian somewhere. Make sure this does not happen to you!

Implementing the support vector machine

We already know how to build an SVM in OpenCV, so there's nothing much to see here. Planning ahead, we wrap the training procedure into a function, so that it's easier to repeat the procedure in the future:

```
In [15]: def train_svm(X_train, y_train):
   ...       svm = cv2.ml.SVM_create()
   ...       svm.train(X_train, cv2.ml.ROW_SAMPLE, y_train)
   ...       return svm
```

The same can be done for the scoring function. Here we pass a feature matrix X and a label vector y, but we do not specify whether we're talking about the training or the test set. In fact, from the viewpoint of the function, it doesn't matter what set the data samples belong to, as long as they have the right format:

```
In [16]: def score_svm(svm, X, y):
   ...       from sklearn import metrics
   ...       _, y_pred = svm.predict(X)
   ...       return metrics.accuracy_score(y, y_pred)
```

Then we can train and score the SVM with two short function calls:

```
In [17]: svm = train_svm(X_train, y_train)
In [18]: score_svm(svm, X_train, y_train)
Out[18]: 1.0
In [19]: score_svm(svm, X_test, y_test)
Out[19]: 0.64615384615384619
```

Thanks to the HOG feature descriptor, we make no mistake on the training set. However, our generalization performance is quite abysmal (64.6 percent), as it is much less than the training performance (100 percent). This is an indication that the model is **overfitting** the data. The fact that it is performing way better on the training set than the test set means that the model has resorted to memorizing the training samples, rather than trying to abstract it into a meaningful decision rule. What can we do to improve the model performance?

Bootstrapping the model

An interesting way to improve the performance of our model is to use **bootstrapping**. This idea was actually applied in one of the first papers on using SVMs in combination with HOG features for pedestrian detection. So let's pay a little tribute to the pioneers and try to understand what they did.

Their idea was quite simple. After training the SVM on the training set, they scored the model and found that the model produced some false positives. Remember that false positive means that the model predicted a positive (+) for a sample that was really a negative (-). In our context, this would mean the SVM falsely believed an image to contain a pedestrian. If this happens for a particular image in the dataset, this example is clearly troublesome. Hence, we should add it to the training set and retrain the SVM with the additional troublemaker, so that the algorithm can learn to classify that one correctly. This procedure can be repeated until the SVM gives satisfactory performance.

> We will talk about bootstrapping more formally in Chapter 11, *Selecting the Right Model with Hyperparameter Tuning*.

Let's do the same. We will repeat the training procedure a maximum of three times. After each iteration, we identify the false positives in the test set and add them to the training set for the next iteration. We can break this up into several steps:

1. Train and score the model:

```
In [20]: score_train = []
    ...  score_test = []
    ...  for j in range(3):
    ...      svm = train_svm(X_train, y_train)
    ...      score_train.append(score_svm(svm, X_train, y_train))
    ...      score_test.append(score_svm(svm, X_test, y_test))
```

2. Find the false positives in the test set. If there aren't any, we're done:

```
    ...          _, y_pred = svm.predict(X_test)
    ...          false_pos = np.logical_and((y_test.ravel() == -1),
    ...                                     (y_pred.ravel() == 1))
    ...          if not np.any(false_pos):
    ...              print('no more false positives: done')
    ...              break
```

3. Append the false positives to the training set, and then repeat the procedure:

```
        ...           X_train = np.concatenate((X_train,
        ...                                     X_test[false_pos, :]),
        ...                                    axis=0)
        ...           y_train = np.concatenate((y_train, y_test[false_pos]),
        ...                                    axis=0)
```

That allows us to improve the model over time:

```
In [21]: score_train
Out[21]: [1.0, 1.0]
In [22]: score_test
Out[22]: [0.64615384615384619, 1.0]
```

Here, we achieved 64.6 percent accuracy in the first round, but were able to get that up to a perfect 100 percent in the second round.

You can find the original paper on ResearchGate at https://www.researchgate.net/publication/3703226_Pedestrian_detection_using_wavelet_templates. The paper was presented at the IEEE Computer Society Conference on Computer Vision and Pattern Recognition (CVPR) in 1997 by M. Oren, P. Sinha, and T. Poggio from MIT, doi: 10.1109/CVPR.1997.609319.

Detecting pedestrians in a larger image

What's left to do is to connect the SVM classification procedure with the process of detection. The way to do this is to repeat our classification for every possible patch in the image. This is similar to what we did earlier when we visualized decision boundaries; we created a fine grid and classified every point on that grid. The same idea applies here. We divide the image into patches and classify every patch as either containing a pedestrian or not.

Therefore, if we want to do this, we have to loop over all possible patches in an image, each time shifting our region of interest by a small number of `stride` pixels:

```
In [23]: stride = 16
    ...:     found = []
    ...:     for ystart in np.arange(0, img_test.shape[0], stride):
    ...:         for xstart in np.arange(0, img_test.shape[1], stride):
```

We want to make sure that we do not go beyond the image boundaries:

```
    ...:             if ystart + hroi > img_test.shape[0]:
    ...:                 continue
    ...:             if xstart + wroi > img_test.shape[1]:
```

Detecting Pedestrians with Support Vector Machines

```
    ...                 continue
```

Then we cut out the region of interest, preprocess it, and classify it:

```
    ...             roi = img_test[ystart:ystart + hroi,
    ...                            xstart:xstart + wroi, :]
    ...             feat = np.array([hog.compute(roi, (64, 64))])
    ...             _, ypred = svm.predict(feat)
```

If that particular patch happens to be classified as a pedestrian, we add it to the list of successes:

```
    ...             if np.allclose(ypred, 1):
    ...                 found.append((ystart, xstart, hroi, wroi))
```

Because pedestrians could appear not just at various locations but also in various sizes, we would have to rescale the image and repeat the whole process. Thankfully, OpenCV has a convenience function for this **multi-scale detection task** in the form of the `detectMultiScale` function. This is a bit of a hack, but we can pass all SVM parameters to the `hog` object:

```
In [24]: rho, _, _ = svm.getDecisionFunction(0)
    ...  sv = svm.getSupportVectors()
    ...  hog.setSVMDetector(np.append(sv.ravel(), rho))
```

Then it's possible to call the detection function:

```
In [25]: found = hog.detectMultiScale(img_test)
```

The function will return a list of bounding boxes that contain detected pedestrians.

This seems to work only for linear SVM classifiers. The OpenCV documentation is terribly inconsistent across versions in this regard, so I'm not sure at which version this started or stopped working. Be careful!

In practice, when people are faced with a standard task such as pedestrian detection, they often rely on precanned SVM classifiers that are built into OpenCV. This is the method that I hinted at in the very beginning of this chapter. By loading either `cv2.HOGDescriptor_getDaimlerPeopleDetector()` or `cv2.HOGDescriptor_getDefaultPeopleDetector()`, we can get started with only a few lines of code:

```
In [26]: hogdef = cv2.HOGDescriptor()
    ...  pdetect = cv2.HOGDescriptor_getDefaultPeopleDetector()
In [27]: hogdef.setSVMDetector(pdetect)
```

```
In [28]: found, _ = hogdef.detectMultiScale(img_test)
```

It's easy to plot the test image with Matplotlib:

```
In [29]: from matplotlib import patches
    ...  fig = plt.figure()
    ...  ax = fig.add_subplot(111)
    ...  ax.imshow(cv2.cvtColor(img_test, cv2.COLOR_BGR2RGB))
```

Then we can mark the detected pedestrians in the image by looping over the bounding boxes in `found`:

```
    ...  for f in found:
    ...      ax.add_patch(patches.Rectangle((f[0], f[1]), f[2], f[3],
    ...                                     color='y', linewidth=3,
    ...                                     fill=False))
```

The result looks like this:

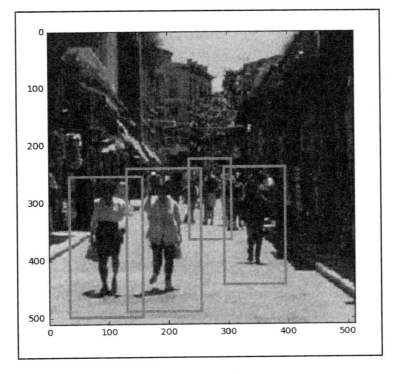

Detected pedestrians in a test image

Further improving the model

Although the RBF kernel makes for a good default kernel, it is not always the one that works best for our problem. The only real way to know which kernel works best on our data is to try them all and compare the classification performance across models. There are strategic ways to perform this so called **hyperparameter tuning**, which we'll talk about in detail in `Chapter 11`, *Selecting the Right Model with Hyper-Parameter Tuning*.

What if we don't know how to do hyperparameter tuning properly yet?

Well, I'm sure you remember the first step in data understanding, *visualize the data*. Visualizing the data could help us understand if a linear SVM were powerful enough to classify the data, in which case there would be no need for a more complex model. After all, we know that if we can draw a straight line to separate the data, a linear classifier is all we need. However, for a more complex problem, we would have to think harder about what shape the optimal decision boundary should have.

In more general terms, we should think about the **geometry** of our problem. Some questions we might ask ourselves are:

- Does visualizing the dataset make it obvious which kind of decision rule would work best? For example, is a straight line good enough (in which case, we would use a linear SVM)? Can we group different data points into blobs or hotspots (in which case, we would use an RBF kernel)?
- Is there a family of data transformations that do not fundamentally change our problem? For example, could we flip the plot on its head, or rotate it, and get the same result? Our kernel should reflect that.
- Is there a way to preprocess our data that we haven't tried yet? Spending some time on feature engineering can make the classification problem much easier in the end. Perhaps we might even be able to use a linear SVM on the transformed data.

As a follow-up exercise, think about which kernel we would use to classify the following datasets:

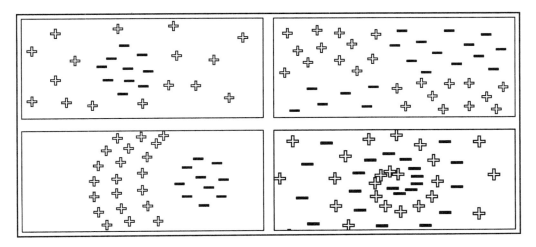

Different datasets that can be classified with SVMs

Think you know the answer? Code it up!

You can use the pregenerated datasets in the Jupyter Notebook `notebooks/06.03-Additional-SVM-Exercises.ipynb` on GitHub. There you will also find the right answers.

Summary

In this chapter, we learned about SVMs in all their forms and flavors. We now know how to draw decision boundaries in 2D and hyperplanes in high-dimensional spaces. We learned about different SVM kernels, and got to know how to implement them in OpenCV.

In addition, we also applied our newly gained knowledge to the practical example of pedestrian detection. For this, we had to learn about the HOG feature descriptor, and how to collect suitable data for the task. We used bootstrapping to improve the performance of our classifier and combined the classifier with OpenCV's multi-scale detection mechanism.

Not only was that a lot to digest in a single chapter, but you have also made it through half of the book. Congrats!

In the next chapter, we will shift gears a bit and revisit a topic from earlier chapters: spam filters. However, this time we want to build a much smarter one than in the first chapter, using the power of Bayesian decision theory.

Implementing a Spam Filter with Bayesian Learning

Before we get to grips with advanced topics, such as cluster analysis, deep learning, and ensemble models, let's turn our attention to a much simpler model that we have overlooked so far: the **naive Bayes classifier**.

Naive Bayes classifiers have their roots in **Bayesian inference**, named after famed statistician and philosopher Thomas Bayes (1701-1761). **Bayes' theorem** famously describes the **probability** of an event based on **prior knowledge** of conditions that might lead to the event. We can use Bayes' theorem to build a statistical model that can not only classify data but also provide us with an estimate of how likely it is that our classification is correct. In our case, we can use Bayesian inference to dismiss an email as spam with high confidence, and to determine the probability of a woman having breast cancer, given a positive screening test.

We have now gained enough experience with the mechanics of implementing machine learning methods, and so we should no longer be afraid to try and understand the theory behind them. Don't worry, we won't write a book on it (ahem), but we need some understanding of the theory to appreciate a model's inner workings. After that, I am sure you will find that Bayesian classifiers are easy to implement, computationally efficient, and tend to perform quite well on relatively small datasets.

Along the way, we also want to address the following questions:

- What makes a naive Bayes classifier so darn naive in the first place?
- What is the difference between discriminative and generative models?
- How do I implement Bayesian classifiers in OpenCV and how do I use them to detect fraudulent email messages?

Understanding Bayesian inference

Although Bayesian classifiers are relatively simple to implement, the theory behind them can be quite counter-intuitive at first, especially if you are not too familiar with **probability theory** yet. However, the beauty of Bayesian classifiers is that they understand the underlying data better than all the classifiers we have encountered so far. For example, standard classifiers, such as the *k*-nearest neighbor algorithm or decision trees, might be able to tell us the target label of a never-before-seen data point. However, these algorithms have no concept of how likely it is for their predictions to be right or wrong. We call them **discriminative models**. Bayesian models, on the other hand, have an understanding of the underlying **probability distribution** that caused the data. We call them **generative models** because they don't just put labels on existing data points- they can also generate new data points with the same statistics.

If this last paragraph was a bit over your head, you might enjoy the following briefer on probability theory. It will be important for the upcoming sections.

Taking a short detour on probability theory

In order to appreciate Bayes' theorem, we need to get a hold of the following technical terms:

- **Random variable**: This is a variable whose value depends on chance. A good example is the act of flipping a coin, which might turn up heads or tails. If a random variable can take on only a limited number of values, we call it **discrete** (such as a coin flip or a dice roll), otherwise we call it a **continuous** random variable (such as the temperature on a given day). Random variables are often typeset as capital letters.
- **Probability**: This is a measure of how likely it is for an event to occur. We denote the probability of an event *e* happening as *p(e)*, which must be a number between 0 and 1 (or between 0 and 100%). For example, for the random variable X denoting a coin toss, we can describe the probability of getting heads as *p(X = heads) = 0.5* (if the coin is fair).
- **Conditional probability**: This is the measure of the probability of an event, given that another event has occurred. We denote the probability of event *y* happening, given that we know event *x* has already happened, as *p(y|x)* (read *p of y given x*). For example, the probability of tomorrow being Tuesday if today is Monday is *p(tomorrow will be Tuesday | today is Monday) = 1*. Of course, there's also the chance that we won't see a tomorrow, but let's ignore that for now.

- **Probability distribution**: This is a function that tells you the probability of different events happening in an experiment. For discrete random variables, this function is also called the **probability mass function**. For continuous random variables, this function is also called the **probability density function**. For example, the probability distribution of the coin toss X would take the value 0.5 for X = *heads*, and 0.5 for X = *tails*. Across all possible outcomes, the distribution must add up to 1.
- **Joint probability distribution**: This is basically the preceding probability function applied to multiple random variables. For example, when flipping two fair coins A and B, the joint probability distribution would enumerate all possible outcomes—(A = *heads*, B = *heads*), (A = *heads*, B = *tails*), (A = *tails*, B = *heads*), and (A = *tails*, B = *tails*)—and tell you the probability of each of these outcomes.

Now here's the trick: If we think of a dataset as a random variable X, then a machine learning model is basically trying to learn the mapping of X onto a set of possible target labels Y. In other words, we are trying to learn the conditional probability $p(Y|X)$, which is the probability that a random sample drawn from X has a target label Y.

There are two different ways to learn $p(Y|X)$, as briefly mentioned earlier:

- **Discriminative models**: This class of models directly learns $p(Y|X)$ from training data without wasting any time trying to understand the underlying probability distributions (such as $p(X)$, $p(Y)$, or even $p(Y|X)$). This approach applies to pretty much every learning algorithm we have encountered so far: linear regression, k-nearest neighbors, decision trees, and so on.
- **Generative models**: This class of model learns everything about the underlying probability distribution and then infers $p(Y|X)$ from the joint probability distribution $p(X, Y)$. Because these models know $p(X, Y)$, they can not only tell us how likely it is for a data point to have a particular target label but also generate entirely new data points. Bayesian models are one example of this class of models.

Cool, so how do Bayesian models actually work? Let's have a look at a specific example.

Understanding Bayes' theorem

There are quite a few scenarios where it would be really good to know how likely it is for our classifier to make a mistake. For example, in Chapter 5, *Using Decision Trees to Make a Medical Diagnosis*, we trained a decision tree to diagnose women with breast cancer based on some medical tests. You can imagine that in this case, we would want to avoid a misdiagnosis at all costs; diagnosing a healthy woman with breast cancer (a false positive) would be both soul-crushing and lead to unnecessary, expensive medical procedures, whereas missing a woman's breast cancer (a false negative) might eventually cost the woman her life.

Good to know we have Bayesian models to count on. Let's walk through a specific (and quite famous) example from http://yudkowsky.net/rational/bayes:

> *1% of women at age forty who participate in routine screening have breast cancer. 80% of women with breast cancer will get positive mammographies. 9.6% of women without breast cancer will also get positive mammographies. A woman in this age group had a positive mammography in a routine screening. What is the probability that she actually has breast cancer?*

What do you think the answer is?

Well, given that her mammography was positive, you might reason that the probability of her having cancer is quite high (somewhere near 80%). It seems much less likely that the woman would belong to the 9.6% with false positives, so the real probability is probably somewhere between 70% and 80%.

I'm afraid that's not correct.

Here's one way to think about this problem. For the sake of simplicity, let's assume we're looking at some concrete number of patients, say 10,000. Before the mammography screening, the 10,000 women can be divided into two groups:

- **Group X**: 100 women *with* breast cancer
- **Group Y**: 9,900 women *without* breast cancer

So far, so good. If we sum up the numbers in the two groups, we get a total of 10,000 patients, confirming that nobody has been lost in the math. After the mammography screening, we can divide the 10,000 women into four groups:

- **Group 1**: 80 women with breast cancer, and a positive mammography
- **Group 2**: 20 women with breast cancer, and a negative mammography

- **Group 3**: 950 women without breast cancer, and a positive mammography
- **Group 4**: 8,950 women without breast cancer, and a negative mammography

As you can see, the sum of all the four groups is still 10,000. The sum of Groups 1 and 2 (the groups with breast cancer) corresponds to Group X, and the sum of Groups 3 and 4 (the groups without breast cancer) corresponds to Group Y. This might become clearer when we draw it out:

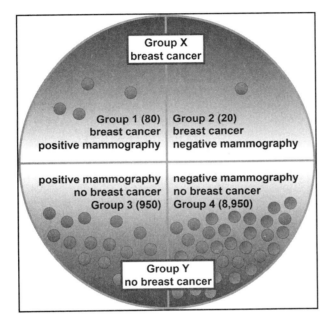

Dividing the data points into groups

In this figure, the top half corresponds to Group X, and the bottom half corresponds to Group Y. Analogously, the left half corresponds to all women with positive mammographies, and the right half corresponds to all women with negative mammographies.

Now, it is easier to see that what we are looking for concerns only the left half of the figure. The proportion of cancer patients with positive results within the group of all patients with positive results is the proportion of Group 1 within Groups 1 and 3:

$$80 / (80 + 950) = 80 / 1{,}030 = 7.8\%$$

In other words, if you administer a mammography to 10,000 patients, then out of the 1,030 with positive mammographies, 80 of those positive mammography patients will have cancer. This is the correct answer, the answer a doctor should give a positive mammography patient if she asks about the chance she has breast cancer; if thirteen patients ask this question, roughly 1 out of those 13 will have cancer.

What we just calculated is called a **conditional probability**: What is our **degree of belief** that a woman has breast cancer **under the condition** (we also say **given**) a positive mammography? As in the last subsection, we denote this with $p(cancer \mid mammography)$, or $p(C \mid M)$, for short. Using capital letters once again enforces the idea that both health and the mammography can have several outcomes, depending on several underlying (and possibly unknown) causes. Therefore, they are random variables.

Then, we can express $P(C \mid M)$ with the following formula:

$$p(C \mid M) = \frac{p(C, M)}{p(C, M) + p(\sim C, M)}.$$

Here, $p(C, M)$ denotes the probability that both C and M are true (meaning the probability that a woman both has cancer and a positive mammography). This is equivalent to the probability of a woman belonging to Group 1 as shown earlier.

The comma (,) means logical *and*, and the tilde (~) stands for logical *not*. Thus, $p(\sim C, M)$ denotes the probability that C is not true and M is true (meaning the probability that a woman does not have cancer but has a positive mammography). This is equivalent to the probability of a woman belonging to Group 3. So, the denominator basically adds up women in Group 1 ($p(C, M)$) and Group 3 ($p(\sim C, M)$).

But wait! Those two groups together simply denote the probability of a woman having a positive mammography, $p(M)$. Therefore, we can simplify the preceding equation:

$$p(C \mid M) = \frac{p(C, M)}{p(M)}.$$

The Bayesian version is to re-interpret what $p(C, M)$ means. We can express $p(C, M)$ as follows:

$$p(C, M) = p(M \mid C) p(C).$$

Now it gets a bit confusing. Here, *p(C)* is simply the probability that a woman has cancer (corresponding to the aforementioned Group X). Given that a woman has cancer, what is the probability that her mammography will be positive? From the problem question, we know it is 80%. This is *p(M | C)*, the probability of *M*, given *C*.

Replacing *p(C, M)* in the first equation with this new formula, we get the following equation:

$$p(C \mid M) = \frac{p(C) p(M \mid C)}{p(M)}$$

In the Bayesian world, these terms all have their specific names:

- *p(C | M)* is called the **posterior**, which is always the thing we want to compute. In our example, this corresponds to the degree of belief that a woman has breast cancer, given a positive mammography.
- *p(C)* is called the **prior** as it corresponds to our initial knowledge about how common breast cancer is. We also call this our initial degree of belief in *C*.
- *p(M | C)* is called the **likelihood**.
- *p(M)* is called the **evidence**.

So, you can rewrite the equation one more time as follows:

$$\text{posterior} = \frac{\text{prior} \times \text{likelihood}}{\text{evidence}}$$

In practice, there is interest only in the numerator of that fraction because the denominator does not depend on *C*, so the denominator is basically constant and can be neglected.

Understanding the naive Bayes classifier

So far, we have only talked about one piece of evidence. However, in most real-world scenarios, we have to predict an outcome (such as a random variable Y) given multiple pieces of evidence (such as random variables X_1 and X_2). So, instead of calculating $p(Y|X)$, we would often have to calculate $p(Y|X_1, X_2, ..., X_n)$. Unfortunately, this makes the math very complicated. For two random variables X_1 and X_2, the joint probability would be computed like this:

$$p(C, X_1, X_2) = p(X_1 | X_2, C) p(X_2 | C) p(C)$$

The ugly part is the term $p(X_1 | X_2, C)$, which says that the conditional probability of X_1 depends on all other variables, including C. This gets even uglier in the case of n variables $X_1, X_2, ..., X_n$.

Thus, the idea of the naive Bayes classifier is simply to ignore all those mixed terms and instead assume that all the pieces of evidence (X_i) are independent of each other. Since this is rarely the case in the real world, you might call this assumption a bit **naive**. And that's where the naive Bayes classifier got its name from.

The assumption that all the pieces of evidence are independent simplifies the term $p(X_1 | X_2, C)$ to $p(X_1 | C)$. In the case of two pieces of evidence (X_1 and X_2), the last equation simplifies to the following equation:

$$p(C | X_1, X_2) = p(X_1 | C) p(X_2 | C) p(C)$$

This simplifies our lives a lot, since we know how to calculate all the terms in the equation. More generally, for n pieces of evidence, we get the following equation:

$$p(C_1 | X_1, X_2, ..., X_n) = p(X_1 | C_1) p(X_2 | C_1) ... p(X_n | C_1) p(C_1)$$

This is the probability that we would predict class C_1, given $X_1, ..., X_n$. We could write a second equation that would do the same for another class, C_2, given $X_1, ..., X_n$, and for a third and a fourth. This is the secret sauce of the naive Bayes classifier.

The naive Bayes classifier then combines this model with a decision rule. The most common rule is to check all the equations (for all classes $C_1, C_2, ..., C_m$) and then pick the class that has the highest probability. This is also known as the **maximum a posteriori** (**MAP**) decision rule.

Implementing your first Bayesian classifier

But enough with the math, let's do some coding!

In the previous chapter, we learned how to generate a number of Gaussian blobs using scikit-learn. Do you remember how that is done?

Creating a toy dataset

The function I'm referring to resides within scikit-learn's `datasets` module. Let's create 100 data points, each belonging to one of two possible classes, and group them into two Gaussian blobs. To make the experiment reproducible, we specify an integer to pick a seed for the `random_state`. You can again pick whatever number you prefer. Here I went with Thomas Bayes' year of birth (just for kicks):

```
In [1]: from sklearn import datasets
   ...     X, y = datasets.make_blobs(100, 2, centers=2,
           random_state=1701, cluster_std=2)
```

Let's have a look at the dataset we just created using our trusty friend, Matplotlib:

```
In [2]: import matplotlib.pyplot as plt
   ...     plt.style.use('ggplot')
   ...     %matplotlib inline
In [3]: plt.scatter(X[:, 0], X[:, 1], c=y, s=50);
```

I'm sure this is getting easier every time. We use `scatter` to create a scatter plot of all *x* values (`X[:, 0]`) and *y* values (`X[:, 1]`), which will result in the following output:

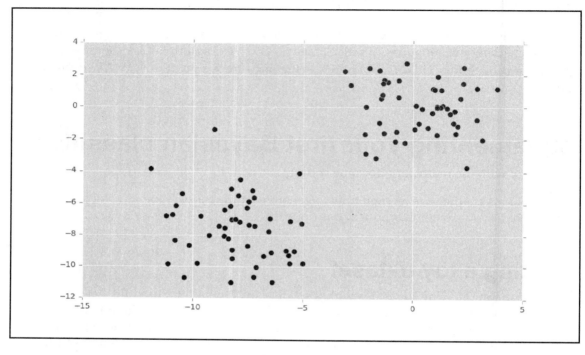

Scatter plot of the Thomas Bayes-seeded dataset

In agreement with our specifications, we see two different point clusters. They hardly overlap, so it should be relatively easy to classify them. What do you think—could a linear classifier do the job?

Yes, it could. Recall that a linear classifier would try to draw a straight line through the figure, trying to put all blue dots on one side and all red dots on the other. A diagonal line going from the top-left corner to the bottom-right corner could clearly do the job. So we would expect the classification task to be relatively easy, even for a naive Bayes classifier.

But first, don't forget to split the dataset into training and test sets! Here, I reserve 10% of the data points for testing:

```
In [4]: import numpy as np
   ...: from sklearn import model_selection as ms
   ...: X = X.astype(np.float32)
   ...: X_train, X_test, y_train, y_test = ms.train_test_split(
   ...:     X, y, test_size=0.1
```

...   )

Classifying the data with a normal Bayes classifier

We will then use the same procedure as in earlier chapters to train a **normal Bayes classifier**. Wait, why not a naive Bayes classifier? Well, it turns out OpenCV doesn't really provide a true naive Bayes classifier. Instead, it comes with a Bayesian classifier that doesn't necessarily expect features to be independent, but rather expects the data to be clustered into Gaussian blobs. This is exactly the kind of dataset we created earlier!

We can create a new classifier using the following function:

```
In [5]: import cv2
   ...: model_norm = cv2.ml.NormalBayesClassifier_create()
```

Then, training is done via the `train` method:

```
In [6]: model_norm.train(X_train, cv2.ml.ROW_SAMPLE, y_train)
Out[6]: True
```

Once the classifier has been trained successfully, it will return `True`. We go through the motions of predicting and scoring the classifier, just like we have done a million times before:

```
In [7]: _, y_pred = model_norm.predict(X_test)
In [8]: from sklearn import metrics
   ...: metrics.accuracy_score(y_test, y_pred)
Out[8]: 1.0
```

Even better—we can reuse the plotting function from the last chapter to inspect the decision boundary! If you recall, the idea was to create a mesh grid that would encompass all data points and then classify every point on the grid. The mesh grid is created via the NumPy function of the same name:

```
In [9]: def plot_decision_boundary(model, X_test, y_test):
   ...:     # create a mesh to plot in
   ...:     h = 0.02 # step size in mesh
   ...:     x_min, x_max = X_test[:, 0].min() - 1, X_test[:, 0].max() + 1
   ...:     y_min, y_max = X_test[:, 1].min() - 1, X_test[:, 1].max() + 1
   ...:     xx, yy = np.meshgrid(np.arange(x_min, x_max, h),
   ...:                          np.arange(y_min, y_max, h))
```

The `meshgrid` function will return two floating-point matrices `xx` and `yy` that contain the *x* and *y* coordinates of every coordinate point on the grid. We can flatten these matrices into column vectors using the `ravel` function, and stack them to form a new matrix `X_hypo`:

```
...         X_hypo = np.column_stack((xx.ravel().astype(np.float32),
...                                   yy.ravel().astype(np.float32)))
```

`X_hypo` now contains all *x* values in `X_hypo[:, 0]` and all *y* values in `X_hypo[:, 1]`. This is a format that the `predict` function can understand:

```
...         ret = model.predict(X_hypo)
```

However, we want to be able to use models from both OpenCV and scikit-learn. The difference between the two is that OpenCV returns multiple variables (a Boolean flag indicating success/failure and the predicted target labels), whereas scikit-learn returns only the predicted target labels. Hence, we can check whether the `ret` output is a tuple, in which case we know we're dealing with OpenCV. In this case, we store the second element of the tuple (`ret[1]`). Otherwise, we are dealing with scikit-learn and don't need to index into `ret`:

```
...         if isinstance(ret, tuple):
...             zz = ret[1]
...         else:
...             zz = ret
...         zz = zz.reshape(xx.shape)
```

All that's left to do is to create a contour plot where `zz` indicates the color of every point on the grid. On top of that, we plot the data points using our trusty scatter plot:

```
...         plt.contourf(xx, yy, zz, cmap=plt.cm.coolwarm, alpha=0.8)
...         plt.scatter(X_test[:, 0], X_test[:, 1], c=y_test, s=200)
```

We call the function by passing a model (`model_norm`), a feature matrix (`X`), and a target label vector (`y`):

```
In [10]: plot_decision_boundary(model_norm, X, y)
```

The output looks like this:

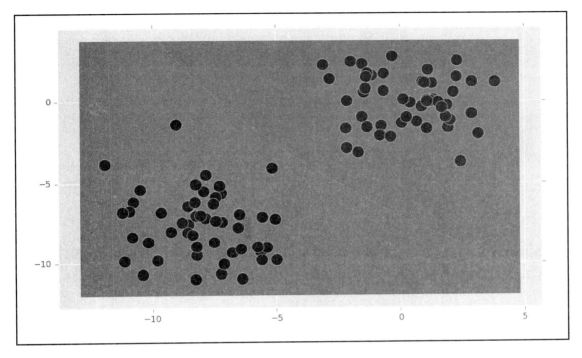

Decision boundary of a normal Bayes classifier

So far, so good. The interesting part is that a Bayesian classifier also returns the probability with which each data point has been classified:

```
In [11]: ret, y_pred, y_proba = model_norm.predictProb(X_test)
```

The function returns a Boolean flag (`True` for success, `False` for failure), the predicted target labels (`y_pred`), and the conditional probabilities (`y_proba`). Here, `y_proba` is an $N \times 2$ matrix that indicates, for every one of the N data points, the probability with which it was classified as either class 0 or class 1:

```
In [12]: y_proba.round(2)
Out[12]: array([[ 0.15000001,  0.05      ],
                [ 0.08      ,  0.        ],
                [ 0.        ,  0.27000001],
                [ 0.        ,  0.13      ],
                [ 0.        ,  0.        ],
                [ 0.18000001,  1.88      ],
                [ 0.        ,  0.        ],
                [ 0.        ,  1.88      ],
                [ 0.        ,  0.        ],
                [ 0.        ,  0.        ]], dtype=float32)
```

This means that for the first data point (top row), the probability of it belonging to class 0 (that is, $p(C_0|X)$) is 0.15 (or 15%)). Similarly, the probability of belonging to class 1 is $p(C_1|X) = 0.05$.

The reason why some of the rows show values greater than 1 is that OpenCV does not really return probability values. Probability values are always between 0 and 1, and each row in the preceding matrix should add up to 1. Instead, what is being reported is a **likelihood**, which is basically the numerator of the conditional probability equation, $p(C) p(M | C)$. The denominator, $p(M)$, does not need to be computed. All we need to know is that 0.15 > 0.05 (top row). Hence, the data point most likely belongs to class 0.

Classifying the data with a naive Bayes classifier

We can compare the result to a true naïve Bayes classifier by asking scikit-learn for help:

```
In [13]: from sklearn import naive_bayes
    ...     model_naive = naive_bayes.GaussianNB()
```

As usual, training the classifier is done via the `fit` method:

```
In [14]: model_naive.fit(X_train, y_train)
Out[14]: GaussianNB(priors=None)
```

Scoring the classifier is built in:

```
In [15]: model_naive.score(X_test, y_test)
Out[15]: 1.0
```

Again a perfect score! However, in contrast to OpenCV, this classifier's `predict_proba` method returns true probability values, because all values are between 0 and 1, and because all rows add up to 1:

```
In [16]: yprob = model_naive.predict_proba(X_test)
    ...     yprob.round(2)
Out[16]: array([[ 0.,  1.],
                [ 0.,  1.],
                [ 0.,  1.],
                [ 1.,  0.],
                [ 1.,  0.],
                [ 1.,  0.],
                [ 0.,  1.],
                [ 0.,  1.],
                [ 1.,  0.],
```

```
         [ 1.,    0.]])
```

You might have noticed something else: This classifier has absolutely no doubt about the target label of each and every data point. It's all or nothing.

The decision boundary returned by the naive Bayes classifier looks slightly different, but can be considered identical to the previous command for the purpose of this exercise:

```
In [17]: plot_decision_boundary(model_naive, X, y)
```

The output looks like this:

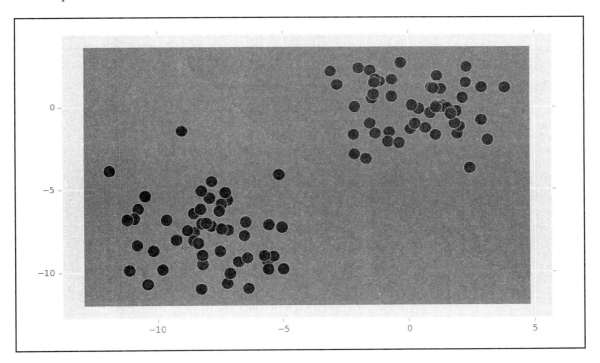

Decision boundary of a naive Bayes classifier

Visualizing conditional probabilities

Similarly, we can also visualize probabilities. For this, we slightly modify the plot function from the previous example. We start out by creating a mesh grid between (x_min, x_max) and (y_min, y_max):

```
In [18]: def plot_proba(model, X_test, y_test):
```

```
...            # create a mesh to plot in
...            h = 0.02 # step size in mesh
...            x_min, x_max = X_test[:, 0].min() - 1, X_test[:, 0].max() + 1
...            y_min, y_max = X_test[:, 1].min() - 1, X_test[:, 1].max() + 1
...            xx, yy = np.meshgrid(np.arange(x_min, x_max, h),
...                                 np.arange(y_min, y_max, h))
```

Then we flatten xx and yy and add them column-wise to the feature matrix X_hypo:

```
...            X_hypo = np.column_stack((xx.ravel().astype(np.float32),
...                                      yy.ravel().astype(np.float32)))
```

If we want to make this function work with both OpenCV and scikit-learn, we need to implement a switch for predictProb (in the case of OpenCV) and predict_proba (in the case of scikit-learn). For this, we check whether model has a method called predictProb. If the method exists, we can call it, otherwise we assume we're dealing with a model from scikit-learn:

```
...            if hasattr(model, 'predictProb'):
...                _, _, y_proba = model.predictProb(X_hypo)
...            else:
...                y_proba = model.predict_proba(X_hypo)
```

Like in In [16] that we saw earlier, y_proba will be a 2D matrix containing, for each data point, the probability of the data belonging to class 0 (in y_proba[:, 0]) and to class 1 (in y_proba[:, 1]). An easy way to convert these two values into a color that the contour function can understand is to simply take the difference of the two probability values:

```
...            zz = y_proba[:, 1] - y_proba[:, 0]
...            zz = zz.reshape(xx.shape)
```

The last step is to plot X_test as a scatter plot on top of the colored meshgrid:

```
...            plt.contourf(xx, yy, zz, cmap=plt.cm.coolwarm, alpha=0.8)
...            plt.scatter(X_test[:, 0], X_test[:, 1], c=y_test, s=200)
```

Now we are ready to call the function:

```
In [19]: plot_proba(model_naive, X, y)
```

The result looks like this:

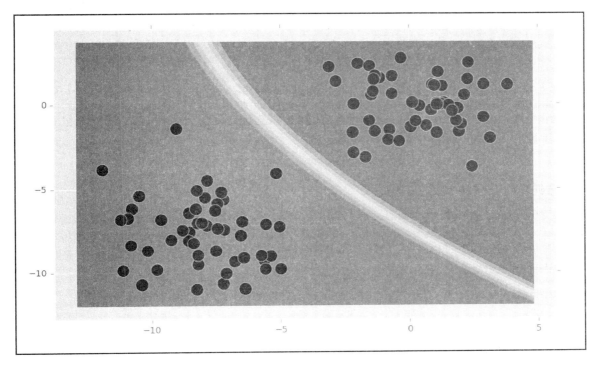

Conditional probabilities of a naive Bayes classifier

Classifying emails using the naive Bayes classifier

The final task of this chapter will be to apply our newly gained skills to a real spam filter!

Naive Bayes classifiers are actually a very popular model for email filtering. Their naivety lends itself nicely to the analysis of text data, where each feature is a word (or a **bag of words**), and it would not be feasible to model the dependence of every word on every other word.

There are a bunch of good email datasets out there, such as the following:

- The Ling-Spam corpus: `http://csmining.org/index.php/ling-spam-datasets.html`
- The Hewlett-Packard spam database: `https://archive.ics.uci.edu/ml/machine-learning-databases/spambase`

- The Enrom-Spam dataset: `http://www.aueb.gr/users/ion/data/enron-spam`
- The Apache SpamAssassin public corpus: `http://csmining.org/index.php/spam-assassin-datasets.html`

In this section, we will be using the Enrom-Spam dataset, which can be downloaded for free from the given website. However, if you followed the installation instructions at the beginning of this book and have downloaded the latest code from GitHub, you are already good to go!

Loading the dataset

If you downloaded the latest code from GitHub, you will find a number of `.zip` files in the `notebooks/data/chapter7` directory. These files contain raw email data (with fields for To:, Cc:, and text body) that are either classified as spam (with the `SPAM = 1` class label) or not (also known as ham, the `HAM = 0` class label).

We build a variable called `sources`, which contains all the raw data files:

```
In [1]: HAM = 0
   ...: SPAM = 1
   ...: datadir = 'data/chapter7'
   ...: sources = [
   ...:     ('beck-s.tar.gz', HAM),
   ...:     ('farmer-d.tar.gz', HAM),
   ...:     ('kaminski-v.tar.gz', HAM),
   ...:     ('kitchen-l.tar.gz', HAM),
   ...:     ('lokay-m.tar.gz', HAM),
   ...:     ('williams-w3.tar.gz', HAM),
   ...:     ('BG.tar.gz', SPAM),
   ...:     ('GP.tar.gz', SPAM),
   ...:     ('SH.tar.gz', SPAM)
   ...: ]
```

The first step is to extract these files into subdirectories. For this, we can use the `extract_tar` function we wrote in the previous chapter:

```
In [2]: def extract_tar(datafile, extractdir):
   ...:     try:
   ...:         import tarfile
   ...:     except ImportError:
   ...:         raise ImportError("You do not have tarfile installed. "
   ...:                           "Try unzipping the file outside of "
   ...:                           "Python.")
   ...:     tar = tarfile.open(datafile)
```

```
    ...            tar.extractall(path=extractdir)
    ...            tar.close()
    ...            print("%s successfully extracted to %s" % (datafile,
    ...                                                       extractdir))
```

In order to apply the function to all data files in the sources, we need to run a loop. The `extract_tar` function expects a path to the `.tar.gz` file—which we build from `datadir` and an entry in `sources`—and a directory to extract the files to (`datadir`). This will extract all emails in, for example, `data/chapter7/beck-s.tar.gz` to the `data/chapter7/beck-s/` subdirectory:

```
In [3]: for source, _ in sources:
   ...:     datafile = '%s/%s' % (datadir, source)
   ...:     extract_tar(datafile, datadir)
Out[3]: data/chapter7/beck-s.tar.gz successfully extracted to data/chapter7
        data/chapter7/farmer-d.tar.gz successfully extracted to
            data/chapter7
        data/chapter7/kaminski-v.tar.gz successfully extracted to
            data/chapter7
        data/chapter7/kitchen-l.tar.gz successfully extracted to
            data/chapter7
        data/chapter7/lokay-m.tar.gz successfully extracted to
            data/chapter7
        data/chapter7/williams-w3.tar.gz successfully extracted to
            data/chapter7
        data/chapter7/BG.tar.gz successfully extracted to data/chapter7
        data/chapter7/GP.tar.gz successfully extracted to data/chapter7
        data/chapter7/SH.tar.gz successfully extracted to data/chapter7
```

Now here's the tricky bit. Every one of these subdirectories contains a number of other directories, wherein the text files reside. So we need to write two functions:

- `read_single_file(filename)`: This is a function that extracts the relevant content from a single file called `filename`
- `read_files(path)`: This is a function that extracts the relevant content from all files in a particular directory called `path`

In order to extract the relevant content from a single file, we need to be aware of how each file is structured. The only thing we know is that the header section of the email (From:, To:, Cc:) and the main body of text are separated by a newline character '\n'. So what we can do is iterate over every line in the text file and keep only those lines that belong to the main text body, which will be stored in the variable lines. We also want to keep a Boolean flag `past_header` around, which is initially set to `False` but will be flipped to `True` once we are past the header section. We start out by initializing those two variables:

```
In [4]: import os
   ...: def read_single_file(filename):
   ...:     past_header, lines = False, []
```

Then we check whether a file with the name `filename` exists. If it does, we start looping over it line by line:

```
   ...:     if os.path.isfile(filename):
   ...:         f = open(filename, encoding="latin-1")
   ...:         for line in f:
```

You may have noticed the `encoding="latin-1"` part. Since some of the emails are not in Unicode, this is an attempt to decode the files correctly.

We do not want to keep the header information, so we keep looping until we encounter the `'\n'` character, at which point we flip `past_header` from `False` to `True`. At this point, the first condition of the following `if-else` clause is met, and we append all remaining lines in the text file to the `lines` variable:

```
   ...:             if past_header:
   ...:                 lines.append(line)
   ...:             elif line == '\n':
   ...:                 past_header = True
   ...:         f.close()
```

In the end, we concatenate all lines into a single string, separated by the newline character, and return both the full path to the file and the actual content of the file:

```
   ...:     content = '\n'.join(lines)
   ...:     return filename, content
```

The job of the second function will be to loop over all files in a folder and call `read_single_file` on them:

```
In [5]: def read_files(path):
   ...:     for root, dirnames, filenames in os.walk(path):
   ...:         for filename in filenames:
   ...:             filepath = os.path.join(root, filename)
   ...:             yield read_single_file(filepath)
```

Here, `yield` is a keyword that is similar to `return`. The difference is that `yield` returns a generator instead of the actual values, which is desirable if you expect to have a large number of items to iterate over.

Building a data matrix using Pandas

Now it's time to introduce another essential data science tool that comes preinstalled with Python Anaconda: **Pandas**. Pandas is built on NumPy and provides a number of useful tools and methods to deal with data structures in Python. Just as we generally import NumPy under the alias `np`, it is common to import Pandas under the `pd` alias:

```
In [6]: import pandas as pd
```

Pandas provides a useful data structure called `DataFrame`, which can be understood as a generalization of a 2D NumPy array, as shown here:

```
In [7]: pd.DataFrame({
   ...          'model': [
   ...              'Normal Bayes',
   ...              'Multinomial Bayes',
   ...              'Bernoulli Bayes'
   ...          ],
   ...          'class': [
   ...              'cv2.ml.NormalBayesClassifier_create()',
   ...              'sklearn.naive_bayes.MultinomialNB()',
   ...              'sklearn.naive_bayes.BernoulliNB()'
   ...          ]
   ...      })
```

The output of the cell will look like this:

Out[7]:

	class	model
0	cv2.ml.NormalBayesClassifier_create()	Normal Bayes
1	sklearn.naive_bayes.MultinomialNB()	Multinomial Bayes
2	sklearn.naive_bayes.BernoulliNB()	Bernoulli Bayes

Example of a Pandas DataFrame

We can combine the preceding functions to build a Pandas DataFrame from the extracted data:

```
In [8]: def build_data_frame(extractdir, classification):
   ...          rows = []
   ...          index = []
   ...          for file_name, text in read_files(extractdir):
```

Implementing a Spam Filter with Bayesian Learning

```
...              rows.append({'text': text, 'class': classification})
...              index.append(file_name)
...
...         data_frame = pd.DataFrame(rows, index=index)
...         return data_frame
```

We then call it with the following command:

```
In [9]: data = pd.DataFrame({'text': [], 'class': []})
...     for source, classification in sources:
...         extractdir = '%s/%s' % (datadir, source[:-7])
...         data = data.append(build_data_frame(extractdir,
                                                classification))
```

Preprocessing the data

Scikit-learn offers a number of options when it comes to encoding text features, which we discussed in Chapter 4, *Representing Data and Engineering Features*. One of the simplest methods of encoding text data, we recall, is by **word count**: For each phrase, you count the number of occurrences of each word within it. In scikit-learn, this is easily done using `CountVectorizer`:

```
In [10]: from sklearn import feature_extraction
...      counts = feature_extraction.text.CountVectorizer()
...      X = counts.fit_transform(data['text'].values)
...      X.shape
Out[10]: (52076, 643270)
```

The result is a giant matrix, which tells us that we harvested a total of 52,076 emails that collectively contain 643,270 different words. However, scikit-learn is smart and saved the data in a sparse matrix:

```
In [11]: X
Out[11]: <52076x643270 sparse matrix of type '<class 'numpy.int64'>'
          with 8607632 stored elements in Compressed Sparse Row
          format>
```

In order to build the vector of target labels (y), we need to access data in the Pandas DataFrame. This can be done by treating the DataFrame like a dictionary, where the `values` attribute will give us access to the underlying NumPy array:

```
In [12]: y = data['class'].values
```

Training a normal Bayes classifier

From here on out, things are (almost) like they always were. We can use scikit-learn to split the data into training and test sets. (Let's reserve 20% of all data points for testing):

```
In [13]: from sklearn import model_selection as ms
    ...: X_train, X_test, y_train, y_test = ms.train_test_split(
    ...:     X, y, test_size=0.2, random_state=42
    ...: )
```

We can instantiate a new normal Bayes classifier with OpenCV:

```
In [14]: import cv2
    ...: model_norm = cv2.ml.NormalBayesClassifier_create()
```

However, OpenCV does not know about sparse matrices (at least its Python interface does not). If we were to pass `X_train` and `y_train` to the `train` function like we did earlier, OpenCV would complain that the data matrix is not a NumPy array. But converting the sparse matrix into a regular NumPy array will likely make you run out of memory. Thus, a possible workaround is to train the OpenCV classifier only on a subset of data points (say 1,000) and features (say 300):

```
In [15]: import numpy as np
    ...: X_train_small = X_train[:1000, :300].toarray().astype(np.float32)
    ...: y_train_small = y_train[:1000].astype(np.float32)
```

Then it becomes possible to train the OpenCV classifier (although this might take a while):

```
In [16]: model_norm.train(X_train_small, cv2.ml.ROW_SAMPLE, y_train_small)
```

Training on the full dataset

However, if you want to classify the full dataset, we need a more sophisticated approach. We turn to scikit-learn's naive Bayes classifier, as it understands how to handle sparse matrices. In fact, if you didn't pay attention and treated `X_train` like every NumPy array before, you might not even notice that anything is different:

```
In [17]: from sklearn import naive_bayes
    ...: model_naive = naive_bayes.MultinomialNB()
    ...: model_naive.fit(X_train, y_train)
Out[17]: MultinomialNB(alpha=1.0, class_prior=None, fit_prior=True)
```

Here we used `MultinomialNB` from the `naive_bayes` module, which is the version of naive Bayes classifier that is best suited to handle categorical data, such as word counts.

The classifier is trained almost instantly and returns the scores for both the training and the test set:

```
In [18]: model_naive.score(X_train, y_train)
Out[18]: 0.95086413826212191
In [19]: model_naive.score(X_test, y_test)
Out[19]: 0.94422043010752688
```

And there we have it: 94.4% accuracy on the test set! Pretty good for not doing much other than using the default values, isn't it?

However, what if we were super critical of our own work and wanted to improve the result even further? There are a couple of things we could do.

Using n-grams to improve the result

One thing to do is to use **n-gram counts** instead of plain word counts. So far, we have relied on what is known as a **bag of words**: We simply threw every word of an email into a bag and counted the number of its occurrences. However, in real emails, the **order** in which **words appear** can carry a great deal of information!

This is exactly what *n*-gram counts are trying to convey. You can think of an *n*-gram as a phrase that is *n* words long. For example, the phrase *Statistics has its moments* contains the following 1-grams: *Statistics*, *has*, *its*, and *moments*. It also has the following 2-grams: *Statistics has*, *has its*, and *its moments*. It also has two 3-grams (*Statistics has its* and *has its moments*), and only a single 4-gram.

We can tell `CountVectorizer` to include any order of *n*-grams into the feature matrix by specifying a range for *n*:

```
In [20]: counts = feature_extraction.text.CountVectorizer(
   ...:         ngram_range=(1, 2)
   ...:     )
   ...: X = counts.fit_transform(data['text'].values)
```

We then repeat the entire procedure of splitting the data and training the classifier:

```
In [21]: X_train, X_test, y_train, y_test = ms.train_test_split(
   ...:         X, y, test_size=0.2, random_state=42
   ...:     )
In [22]: model_naive = naive_bayes.MultinomialNB()
   ...: model_naive.fit(X_train, y_train)
Out[22]: MultinomialNB(alpha=1.0, class_prior=None, fit_prior=True)
```

You might have noticed that the training is taking much longer this time. To our delight, we find that the performance has significantly increased:

```
In [23]: model_naive.score(X_test, y_test)
Out[23]: 0.97062211981566815
```

However, *n*-gram counts are not perfect. They have the disadvantage of unfairly weighting longer documents (because there are more possible combinations of forming *n*-grams). To avoid this problem, we can use **relative frequencies** instead of a simple **number of occurrences**. We have already encountered one way to do so, and it had a horribly complicated name. Do you remember what it was called?

Using tf-idf to improve the result

It was called the **term–inverse document frequency** (tf–idf), and we encountered it in Chapter 4, *Representing Data and Engineering Features*. If you recall, what tf–idf does is basically weigh the word count by a measure of how often they appear in the entire dataset. A useful side effect of this method is the idf part—the inverse frequency of words. This makes sure that frequent words, such as *and*, *the*, and *but*, carry only a small weight in the classification.

We apply tf–idf to the feature matrix by calling `fit_transform` on our existing feature matrix X:

```
In [24]: tfidf = feature_extraction.text.TfidfTransformer()
In [25]: X_new = tfidf.fit_transform(X)
```

Don't forget to split the data:

```
In [26]: X_train, X_test, y_train, y_test = ms.train_test_split(X_new, y,
   ...:         test_size=0.2, random_state=42)
```

Then, when we train and score the classifier again, we suddenly find a remarkable score of 99% accuracy!

```
In [27]: model_naive = naive_bayes.MultinomialNB()
   ...:  model_naive.fit(X_train, y_train)
   ...:  model_naive.score(X_test, y_test)
Out[27]: 0.99087941628264209
```

To convince us of the classifier's awesomeness, we can inspect the **confusion matrix**. This is a matrix that shows, for every class, how many data samples were misclassified as belonging to a different class. The diagonal elements in the matrix tell us how many samples of the class i were correctly classified as belonging to the class i. The off-diagonal elements represent misclassifications:

```
In [28]: metrics.confusion_matrix(y_test, model_naive.predict(X_test))
Out[28]: array([[3746,   84],
                [  11, 6575]])
```

This tells us we got 3,746 class 0 classifications correct, and 6,575 class 1 classifications correct. We confused 84 samples of class 0 as belonging to class 1 and 11 samples of class 1 as belonging to class 0. If you ask me, that's about as good as it gets.

Summary

In this chapter, we took our first stab at probability theory, learning about random variables and conditional probabilities, which allowed us to get a glimpse of Bayes' theorem—the underpinning of a naive Bayes classifier. We talked about the difference between discrete and continuous random variables, between likelihoods and probabilities, between priors and evidence, and between normal and naive Bayes classifiers.

Finally, our theoretical knowledge would be of no use if we didn't apply it to a practical example. We obtained a dataset of raw email messages, parsed it, and trained Bayesian classifiers on it to classify emails as either ham or spam using a variety of feature extraction approaches.

In the next chapter, we will switch gears and, for once, discuss what to do if we have to deal with unlabeled data.

8
Discovering Hidden Structures with Unsupervised Learning

So far, we have focused our attention exclusively on supervised learning problems, where every data point in the dataset had a known label or target value. However, what do we do when there is no known output, or no teacher to supervise the learning algorithm?

This is what **unsupervised learning** is all about. In unsupervised learning, the learning is shown only in the input data and is asked to extract knowledge from this data without further instruction. We have already talked about one of the many forms that unsupervised learning comes in—**dimensionality reduction**. Another popular domain is **cluster analysis**, which aims to partition data into distinct groups of similar items.

In this chapter, we want to understand how different clustering algorithms can be used to extract hidden structures in simple, unlabeled datasets. These hidden structures have many benefits, whether they are used in feature extraction, image processing, or even as a preprocessing step for supervised learning tasks. As a concrete example, we will learn how to apply clustering to images in order to reduce their color spaces to 16 bits.

More specifically, we want to answer the following questions:

- What are *k*-**means clustering** and **expectation-maximization**, and how do I implement these things in OpenCV?
- How can I arrange clustering algorithms in hierarchical trees, and what are the benefits that come from that?
- How can I use unsupervised learning for preprocessing, image processing, and classification?

Let's get started!

Understanding unsupervised learning

Unsupervised learning might come in many shapes and forms, but the goal is always to convert original data into a richer, more meaningful representation, whether that means making it easier for humans to understand or easier for machine learning algorithms to parse.

Some common applications of unsupervised learning include the following:

- **Dimensionality reduction**: This takes a high-dimensional representation of data consisting of many features and tries to compress the data so that its main characteristics can be explained with a small number of highly informative features. For example, when applied to housing prices in the neighborhoods of Boston, dimensionality reduction might be able to tell us that the indicators we should pay most attention to are the property tax and the neighborhood's crime rate.
- **Factor analysis**: This tries to find the hidden causes or unobserved components that gave rise to the observed data. For example, when applied to all the episodes of the 1970s TV show *Scooby-Doo, Where Are You!*, factor analysis might be able to tell us that (spoiler alert!) every ghost or monster on the show is essentially some disgruntled count playing an elaborate hoax on the town.
- **Cluster analysis**: This tries to partition the data into distinct groups of similar items. This is the type of unsupervised learning we will focus on in this chapter. For example, when applied to all the movies on Netflix, cluster analysis might be able to automatically group them into genres.

To make things more complicated, these analyses have to be performed on unlabeled data, where we do not know beforehand what the right answer should be. Consequently, a major challenge in unsupervised learning is to determine whether an algorithm did well or learned anything useful. Often, the only way to evaluate the result of an unsupervised learning algorithm is to inspect it manually and determine by hand whether the result makes sense.

That being said, unsupervised learning can be immensely helpful, for example, as a preprocessing or feature extraction step. You can think of unsupervised learning as a **data transformation**—a way to transform data from its original representation into a more informative form. Learning a new representation might give us deeper insights about our data, and sometimes it might even improve the accuracy of supervised learning algorithms.

Understanding k-means clustering

The most essential clustering algorithm that OpenCV provides is *k*-means clustering, which searches for a predetermined number of *k* clusters (or groups) within an unlabeled multidimensional dataset.

It does so by using two simple assumptions about what an optimal clustering should look like:

- The center of each cluster is simply the arithmetic mean of all the points belonging to the cluster
- Each point in the cluster is closer to its own center than to other cluster centers

It's the easiest to understand the algorithm by looking at a concrete example.

Implementing our first k-means example

First, let's generate a 2D dataset containing four distinct blobs. To emphasize that this is an unsupervised approach, we will leave the labels out of the visualization. We will continue using `matplotlib` for all our visualization purposes:

```
In [1]: import matplotlib.pyplot as plt
   ...   %matplotlib inline
   ...   plt.style.use('ggplot')
```

Following the same recipe from earlier chapters, we will create a total of 300 blobs (`n_samples=300`) belonging to four distinct clusters (`centers=4`):

```
In [2]: from sklearn.datasets.samples_generator import make_blobs
   ...   X, y_true = make_blobs(n_samples=300, centers=4,
   ...                          cluster_std=1.0, random_state=10)
   ...   plt.scatter(X[:, 0], X[:, 1], s=100);
```

This will generate the following figure:

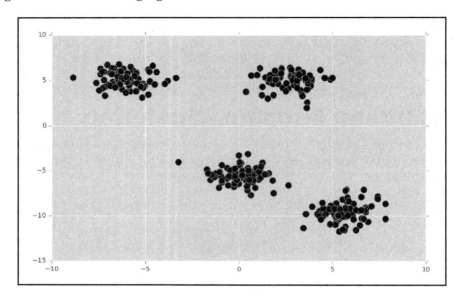

Example dataset of 300 unlabeled points organized into four distinct clusters

Even without assigning target labels to the data, it is straightforward to pick out the four clusters by eye. The *k*-means algorithm can do this, too, without having any information about target labels or underlying data distributions.

Although *k*-means is, of course, a statistical model, in OpenCV, it does not come via the `ml` module and the common `train` and `predict` API calls. Instead, it is directly available as `cv2.kmeans`. In order to use the model, we have to specify some arguments, such as the termination criteria and some initialization flags. Here, we will tell the algorithm to terminate whenever the error is smaller than 1.0 (`cv2.TERM_CRITERIA_EPS`) or when ten iterations have been executed (`cv2.TERM_CRITERIA_MAX_ITER`):

```
In [3]: import cv2
   ...:  criteria = (cv2.TERM_CRITERIA_EPS + cv2.TERM_CRITERIA_MAX_ITER,
   ...:              10, 1.0)
   ...:  flags = cv2.KMEANS_RANDOM_CENTERS
```

Then we can pass the preceding data matrix (X) to `cv2.means`. We also specify the number of clusters (4) and the number of attempts the algorithm should make with different random initial guesses (10), as shown in the following snippet:

```
In [4]: import numpy as np
   ...: compactness, labels, centers = cv2.kmeans(X.astype(np.float32),
   ...:                                            4, None, criteria,
   ...:                                            10, flags)
```

Three different variables are returned. The first one, compactness, returns the sum of squared distances from each point to their corresponding cluster centers. A high compactness score indicates that all points are close to their cluster centers, whereas a low compactness score indicates that the different clusters might not be well separated:

```
In [5]: compactness
Out[5]: 527.01581170992
```

Of course, this number strongly depends on the actual values in X. If the distances between points were large to begin with, we cannot expect an arbitrarily small compactness score. Hence, it is more informative to plot the data points, colored to their assigned cluster labels:

```
In [6]: plt.scatter(X[:, 0], X[:, 1], c=labels, s=50, cmap='viridis')
   ...: plt.scatter(centers[:, 0], centers[:, 1], c='black', s=200,
   ...:             alpha=0.5);
```

This produces a scatter plot of all the data points colored according to which-ever cluster they belong to, with the corresponding cluster centers indicated with a blob of darker shade in the center of every cluster:

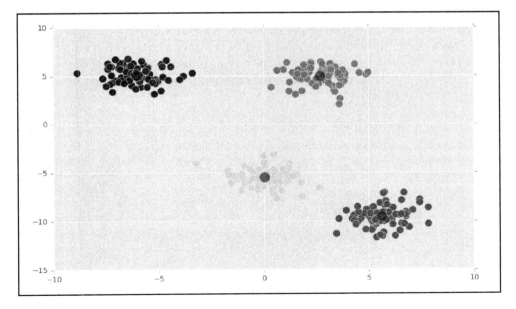

Result of k-means clustering for k = 4

The good news here is that the *k*-means algorithm (at least in this simple case) assigns the points to clusters very similarly to how we might have, had we done the job by eye. But how did the algorithm find these different clusters so quickly? After all, the number of possible combinations of cluster assignments is exponential to the number of data points! By hand, trying all possible combinations would have certainly taken forever.

Fortunately, an exhaustive search is not necessary. Instead, the typical approach that *k*-means takes is to use an iterative algorithm, also known as **expectation-maximization**.

Understanding expectation-maximization

K-means clustering is but one concrete application of a more general algorithm known as expectation-maximization. In short, the algorithm works as follows:

1. Start with some random cluster centers.
2. Repeat until convergence:

    - **Expectation step**: Assign all data points to their nearest cluster center.
    - **Maximization step**: Update the cluster centers by taking the mean of all the points in the cluster.

Here, the expectation step is so named because it involves updating our expectation of which cluster each point in the dataset belongs to. The maximization step is so named because it involves maximizing a fitness function that defines the location of the cluster centers. In the case of *k*-means, maximization is performed by taking the arithmetic mean of all the data points in a cluster.

This should become clearer with the following figure:

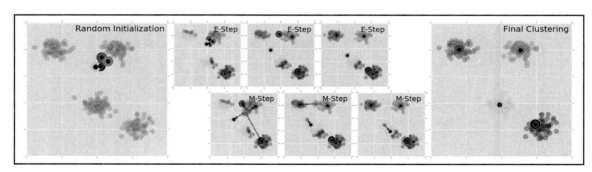

Expectation-maximization step by step

In the preceding image, the algorithm works from left to right. Initially, all data points are grayed out (meaning, we don't know yet which cluster they belong to), and the cluster centers are chosen at random. In each expectation step, the data points are colored according to whichever cluster center they are closest to. In each maximization step, the cluster centers are updated by taking the arithmetic mean of all points belonging to the cluster; the resulting displacement of the cluster centers is indicated with arrows. These two steps are repeated until the **algorithm converges**, meaning until the maximization step does not further improve the clustering outcome.

Another popular application of expectation-maximization is the so—called **Gaussian mixture models (GMMs)**, where the clusters are not spherical but multivariate Gaussians. You can find more information about them at: http://scikit-learn.org/stable/modules/mixture.html.

Implementing our own expectation-maximization solution

The expectation-maximization algorithm is simple enough for us to code it up ourselves. To do so, we will define a function `find_clusters(X, n_clusters, rseed=5)` that takes as input a data matrix (X), the number of clusters we want to discover (n_clusters), and a random seed (optional, `rseed`). As will become clear in a second, scikit-learn's `pairwise_distances_argmin` function will come in handy:

```
In [7]: from sklearn.metrics import pairwise_distances_argmin
   ...:     def find_clusters(X, n_clusters, rseed=5):
```

We can implement expectation-maximization for *k*-means in five essential steps:

1. **Initialization**: Randomly choose a number `n_clusters` of cluster centers. We don't just pick any random number but instead pick actual data points to be the cluster centers. We do this by permuting X along its first axis and picking the first `n_clusters` points in this random permutation:

   ```
   ...:     rng = np.random.RandomState(rseed)
   ...:     i = rng.permutation(X.shape[0])[:n_clusters]
   ...:     centers = X[i]
   ```

2. **While looping forever**: Assign labels based on the closest cluster centers. Here, scikit-learn's `pairwise_distance_argmin` function does exactly what we want: It computes, for each data point in X, the index of the closest cluster center in `centers`:

   ```
   ...         while True:
   ...             labels = pairwise_distances_argmin(X, centers)
   ```

3. **Find the new cluster centers**: In this step, we have to take the arithmetic mean of all data points in X that belong to a specific cluster (`X[labels == i]`):

   ```
   ...             new_centers = np.array([X[labels ==
                   i].mean(axis=0)
   ```

4. **Check for convergence, and break the `while` loop if necessary**: This is the last step to make sure that we stop the execution of the algorithm once the job is done. We determine whether the job is done by checking whether all the new cluster centers are equal to the old cluster centers. If this is true, we exit the loop, otherwise we keep looping:

   ```
   ...             for i in range(n_clusters)])
   ...             if np.all(centers == new_centers):
   ...                 break
   ...             centers = new_centers
   ```

5. Exit the function, and return the result:

   ```
   ...         return centers, labels
   ```

We can apply our function to the preceding data matrix X we created. Since we know what the data looks like, we know that we are looking for four clusters:

```
In [8]: centers, labels = find_clusters(X, 4)
   ...: plt.scatter(X[:, 0], X[:, 1], c=labels, s=100, cmap='viridis');
```

This will generate the following plot:

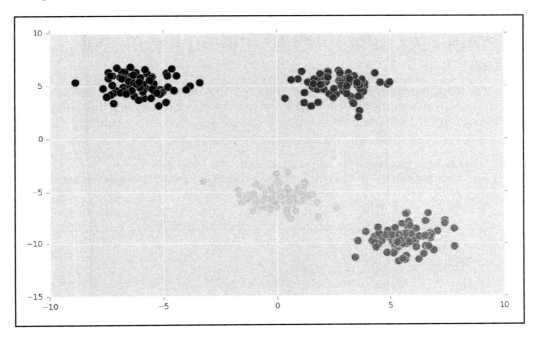

Outcome of our home-made *k*-means using expectation-maximization

As we can see, our home-made algorithm got the job done! Granted, this particular clustering example was fairly easy, and most real-life implementations of *k*-means clustering will do a bit more under the hood. But for now, we are happy.

Knowing the limitations of expectation-maximization

For its simplicity, expectation-maximization performs incredibly well in a range of scenarios. That being said, there are a number of potential limitations that we need to be aware of:

- Expectation-maximization does not guarantee that we will find the globally best solution.
- We must know the number of desired clusters beforehand.
- The decision boundaries of the algorithms are linear.
- The algorithm is slow for large datasets.

Let's quickly discuss these potential caveats in a little more detail.

First caveat: No guarantee of finding the global optimum

Although mathematicians have proved that the expectation-maximization step improves the result in each step, there is still no guarantee that, in the end, we will find the global best solution. For example, if we use a different random seed in our simple example (such as using, seed 10 instead of 5), we suddenly get very poor results:

```
In [9]: centers, labels = find_clusters(X, 4, rseed=10)
   ...      plt.scatter(X[:, 0], X[:, 1], c=labels, s=100, cmap='viridis');
```

This will generate the following plot:

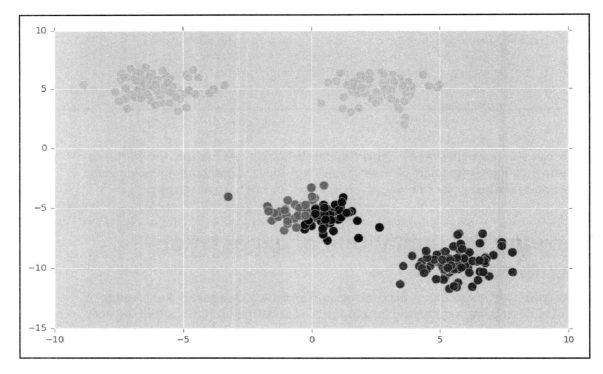

Example of k-means missing the global optimum

What happened?

The short answer is that the random initialization of cluster centers was unfortunate. It led to the center of the yellow cluster migrating in—between the two top blobs, essentially combining them into one. As a result, the other clusters got confused because they suddenly had to split two visually distinct blobs into three clusters.

For this reason, it is common for the algorithm to be run for multiple initial states. Indeed, OpenCV does this by default (set by the optional `attempts` parameter).

Second caveat: We must select the number of clusters beforehand

Another potential limitation is that k-means cannot learn the number of clusters from the data. Instead, we must tell it how many clusters we expect beforehand. You can see how this could be problematic for complicated real-world data that you don't fully understand yet.

From the viewpoint of k-means, there is no wrong or nonsensical number of clusters. For example, if we ask the algorithm to identify six clusters in the dataset generated in the preceding section, it will happily proceed and find the best six clusters:

```
In [10]: criteria = (cv2.TERM_CRITERIA_EPS + cv2.TERM_CRITERIA_MAX_ITER,
   ...                10, 1.0)
   ...   flags = cv2.KMEANS_RANDOM_CENTERS
   ...   compactness, labels, centers = cv2.kmeans(X.astype(np.float32),
   ...                                             6, None, criteria,
   ...                                             10, flags)
```

Here, we used the exact same preceding code and only changed the number of clusters from four to six. We can plot the data points again and color them according to the target labels:

```
In [11]: plt.scatter(X[:, 0], X[:, 1], c=labels, s=100, cmap='viridis');
```

The output looks like this:

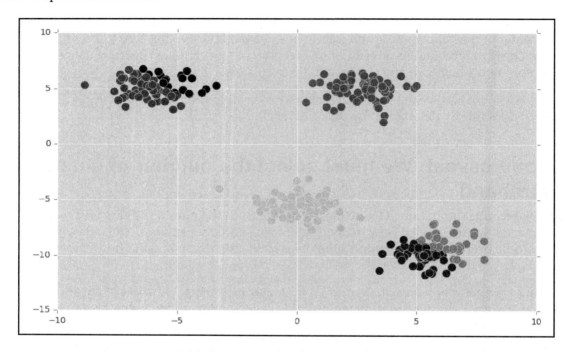

Example of k-means discovering more clusters than there actually are

Since we generated the data ourselves, we know that each data point comes from a total of four distinct clusters. For more complicated data, where we are uncertain about the right number of clusters, there are a few things we could try.

First and foremost, there is the **elbow method**, which asks us to repeat the clustering for a whole range of k values and record the compactness value:

```
In [12]: kvals = np.arange(2, 10)
   ...:   compactness = []
   ...:   for k in kvals:
   ...:       c, _, _ = cv2.kmeans(X.astype(np.float32), k, None,
   ...:                            criteria, 10, flags)
   ...:       compactness.append(c)
```

Then we will plot the compactness as a function of k:

```
In [13]: plt.plot(kvals, compactness, 'o-', linewidth=4,
   ...:           markersize=12)
   ...:  plt.xlabel('number of clusters')
   ...:  plt.ylabel('compactness')
```

This will result in a plot that looks like an *arm*. The point that lies at the *elbow* is the number of clusters to pick:

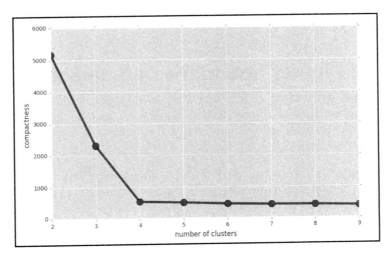

Elbow method

In our case, we would pick the number four. The reasoning behind this is that moving from left to right, four is the smallest number of clusters that gives a very compact representation. Three clusters give us a representation that is only half as compact as four clusters. Choosing five clusters or more does not further improve the compactness. Thus, four should be our best guess as to the number of clusters.

If our results are more complicated than that, we might want to consider a more sophisticated method. Some of the most common ones include the following:

- **Silhouette analysis** allows us to study the separation between the resulting clusters. If a lot of the points in one cluster are closer to a neighboring cluster than their own, we might be better off putting them all in a single cluster.
- **Gaussian mixture models** use something called the Bayesian information criterion to assess the right number of clusters for us.
- **Density-based spatial clustering of applications with noise (DBSCAN)** and **affinity propagation** are two more sophisticated clustering algorithms that can choose a suitable number of clusters for us.

In addition, we can also study the separation distance between the resulting clusters using
what is known as a **silhouette plot**. This plot displays a measure of how close each point in
one cluster is to points in the neighboring clusters, and thus provides a way to assess
parameters, such as the number of clusters, visually. This measure has a range of [-1, 1].

Third caveat: Cluster boundaries are linear

The k-means algorithm is based on a simple assumption, which is that points will be closer
to their own cluster center than to others. Consequently, k-means always assumes linear
boundaries between clusters, meaning that it will fail whenever the geometry of the clusters
is more complicated than that.

We see this limitation for ourselves by generating a slightly more complicated dataset.
Instead of generating data points from Gaussian blobs, we want to organize the data into
two overlapping half circles. We can do this using scikit-learn's `make_moons`. Here, we
choose 200 data points belonging to two half circles, in combination with some Gaussian
noise:

```
In [14]: from sklearn.datasets import make_moons
    ...  X, y = make_moons(200, noise=.05, random_state=12)
```

This time, we tell k-means to look for two clusters:

```
In [15]: criteria = (cv2.TERM_CRITERIA_EPS +
    ...              cv2.TERM_CRITERIA_MAX_ITER, 10, 1.0)
    ...  flags = cv2.KMEANS_RANDOM_CENTERS
    ...  compactness, labels, centers = cv2.kmeans(X.astype(np.float32),
    ...                                            2, None, criteria,
    ...                                            10, flags)
    ...  plt.scatter(X[:, 0], X[:, 1], c=labels, s=100, cmap='viridis');
```

The resulting scatter plot looks like this:

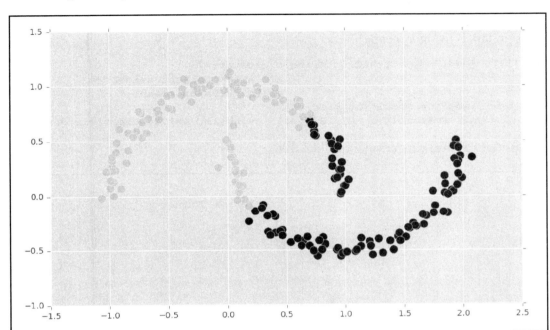

Example of k-means finding linear boundaries in nonlinear data

As is evident from the plot, *k*-means failed to identify the two half circles and instead split the data with what looks like a diagonal straight line (from bottom left to top right).

This scenario should ring a bell. We had the exact same problem when we talked about linear SVMs in Chapter 6, *Detecting Pedestrians with Support Vector Machines*. The idea there was to use the kernel trick in order to transform the data into a higher-dimensional feature space. Can we do the same here?

We most certainly can. There is a form of kernelized *k*-means that works akin to the kernel trick for SVMs, called **spectral clustering**. Unfortunately, OpenCV does not provide an implementation of spectral clustering. Fortunately, scikit-learn does:

```
In [16]: from sklearn.cluster import SpectralClustering
```

The algorithm uses the same API as all other statistical models: We set optional arguments in the constructor and then call `fit_predict` on the data. Here, we want to use the graph of the nearest neighbors to compute a higher-dimensional representation of the data and then assign labels using k-means:

```
In [17]: model = SpectralClustering(n_clusters=2,
   ...:                             affinity='nearest_neighbors',
   ...:                             assign_labels='kmeans')
   ...: labels = model.fit_predict(X)
   ...: plt.scatter(X[:, 0], X[:, 1], c=labels, s=100, cmap='viridis');
```

The output of spectral clustering looks like this:

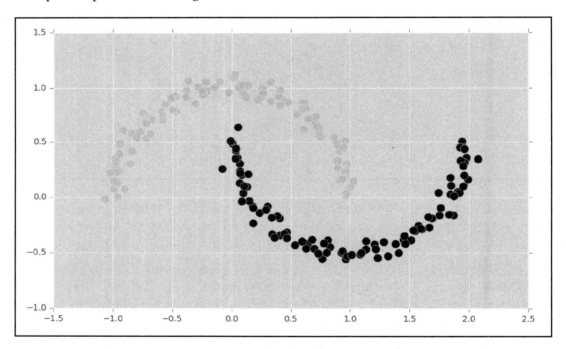

Spectral clustering is able to find nonlinear cluster boundaries

We see that spectral clustering gets the job done. Alternatively, we could have transformed the data into a more suitable representation ourselves and then applied OpenCV's linear k-means to it. The lesson of all this is that perhaps, again, feature engineering saves the day.

Fourth caveat: k-means is slow for a large number of samples

The final limitation of *k*-means is that it is relatively slow for large datasets. You can imagine that quite a lot of algorithms might suffer from this problem. However, k-means is affected especially badly: each iteration of k-means must access every single data point in the dataset and compare it to all the cluster centers.

You might wonder whether the requirement to access all data points during each iteration is really necessary. For example, you might just use a subset of the data to update the cluster centers at each step. Indeed, this is the exact idea that underlies a variation of the algorithm called **batch-based** *k*-means. Unfortunately, this algorithm is not implemented in OpenCV.

K-means is provided by scikit-learn as part of their clustering module: `sklearn.cluster.MiniBatchKMeans`.

Despite the limitations discussed earlier, *k*-means has a number of interesting applications, especially in computer vision.

Compressing color spaces using k-means

One exciting application of *k*-means is the compression of image color spaces. For example, a typical **true-color image** comes with a 24-bit color depth, allowing for a total of 16,777,216 color variations. However, in most images, a large number of the colors will be unused, and many of the pixels in the image will have similar or identical colors.

Alternatively, we can use *k*-means to reduce the color palette to, for example, 16 color variations. The trick here is to think of the cluster centers as the reduced color palette. Then *k*-means will automatically organize the millions of colors in the original image into the appropriate number of colors!

Visualizing the true-color palette

Let's have a look at a particular image:

```
In [1]: import cv2
   ...: import numpy as np
   ...: lena = cv2.imread('data/lena.jpg', cv2.IMREAD_COLOR)
```

By now, we know how to start matplotlib in our sleep:

```
In [2]: import matplotlib.pyplot as plt
   ...     %matplotlib inline
   ...     plt.style.use('ggplot')
```

However, this time we want to disable the grid lines that the `ggplot` option typically displays over images:

```
In [3]: plt.rc('axes', **{'grid': False})
```

Then we can visualize Lena with the following command (don't forget to switch the BGR ordering of the color channels to RGB):

```
In [4]: plt.imshow(cv2.cvtColor(lena, cv2.COLOR_BGR2RGB))
```

This will produce the following output:

A picture of Lena (http://sipi.usc.edu/database/database.php?volume=misc&image=12).

The image itself is stored in a 3D array of size (height, width, depth) containing blue/green/red contributions as integers from 0 to 255:

```
In [5]: lena.shape
Out[5]: (225, 225, 3)
```

Because every color channel has 256 possible values, the number of possible colors is 256 x 256 x 256, or 16,777,216, as mentioned prior to this. One way to visualize the sheer amount of different colors in the image is to reshape the data to a cloud of points in a 3D color space. We also scale the colors to lie between 0 and 1:

```
In [6]: img_data = lena / 255.0
   ...  img_data = img_data.reshape((-1, 3))
   ...  img_data.shape
Out[6]: (50625, 3)
```

In this 3D color space, every row of data is a data point. In order to visualize this data, we will write a function called plot_pixels, which takes as input the data matrix and a figure title. Optionally, we also let the user specify the colors to use. For the sake of efficiency, we can also limit the analysis to a subset of N pixels:

```
In [7]: def plot_pixels(data, title, colors=None, N=10000):
   ...      if colors is None:
   ...          colors = data
   ...
```

If a number N is specified, we will randomly choose N data points from the list of all possible ones:

```
   ...          rng = np.random.RandomState(0)
   ...          i = rng.permutation(data.shape[0])[:N]
   ...          colors = colors[i]
```

Because each data point has three dimensions, plotting is a bit tricky. We could either create a 3D plot using Matplotlib's mplot3d toolkit, or we could produce two plots that visualize a subspace of the 3D point cloud. In this example, we will opt for the latter and generate two plots, one where we plot red versus green and another where we plot red versus blue. For this, we need to access the R, G, and B values of each data point as column vectors:

```
   ...          pixel = data[i].T
   ...          R, G, B = pixel[0], pixel[1], pixel[2]
```

The first plot then shows red on the *x*-axis and green on the *y*-axis:

```
   ...          fig, ax = plt.subplots(1, 2, figsize=(16, 6))
   ...          ax[0].scatter(R, G, color=colors, marker='.')
   ...          ax[0].set(xlabel='Red', ylabel='Green', xlim=(0, 1),
```

```
    ...            ylim=(0, 1))
```

Analogously, the second figure is a scatter plot of red on the *x*-axis and blue on the *y*-axis:

```
    ...         ax[1].scatter(R, B, color=colors, marker='.')
    ...         ax[1].set(xlabel='Red', ylabel='Blue', xlim=(0, 1),
    ...                   ylim=(0, 1))
```

As a last step, we will display the specified title centered over the two subplots:

```
    ...         fig.suptitle(title, size=20);
```

We can then call the function with our data matrix (`data`) and an appropriate title:

```
In [8]: plot_pixels(img_data, title='Input color space: 16 million '
   ...:                            'possible colors')
```

This will produce the following output:

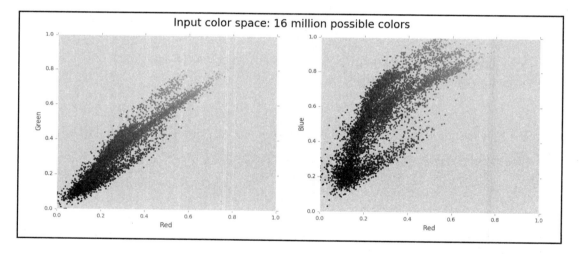

Input color space

Let's see what more we can do with this algorithm.

Reducing the color palette using k-means

Now let's reduce these 16 million colors to just 16 colors by telling *k*-means to cluster all color variations into 16 distinct clusters. We use the before mentioned procedure, but now specify 16 as the number of clusters:

```
In [9]: criteria = (cv2.TERM_CRITERIA_EPS + cv2.TERM_CRITERIA_MAX_ITER,
   ...             10, 1.0)
   ...    flags = cv2.KMEANS_RANDOM_CENTERS
   ...    img_data = img_data.astype(np.float32)
   ...    compactness, labels, centers = cv2.kmeans(img_data,
   ...                                              16, None, criteria,
   ...                                              10, flags)
```

The resulting cluster corresponds to the 16 colors of our reduced color palette. Visual inspection of the `centers` array reveals that all colors have three entries—B, G, and R—with values between 0 and 1:

```
In [10]: centers
Out[10]: array([[ 0.29973754,  0.31500012,  0.48251548],
                [ 0.27192295,  0.35615689,  0.64276862],
                [ 0.17865284,  0.20933454,  0.41286203],
                [ 0.39422086,  0.62827665,  0.94220853],
                [ 0.34117648,  0.58823532,  0.90196079],
                [ 0.42996961,  0.62061119,  0.91163337],
                [ 0.06039202,  0.07102439,  0.1840712 ],
                [ 0.5589878 ,  0.6313886 ,  0.83993536],
                [ 0.37320262,  0.54575169,  0.88888896],
                [ 0.35686275,  0.57385623,  0.88954246],
                [ 0.47058824,  0.48235294,  0.59215689],
                [ 0.34346411,  0.57483661,  0.88627452],
                [ 0.13815609,  0.12984112,  0.21053818],
                [ 0.3752504 ,  0.47029912,  0.75687987],
                [ 0.31909946,  0.54829341,  0.87378371],
                [ 0.40409693,  0.58062142,  0.8547557 ]], dtype=float32)
```

These 16 colors correspond to the 16 cluster labels contained in the `labels` vector. So we want all data points with label 0 to be colored according to row 0 in the `centers` array; all data points with label 1 to be colored according to row 1 in the `centers` array; and so on. In other words, we want to use `labels` as an index into the `centers` array—these are our new colors:

```
In [11]: new_colors = centers[labels].reshape((-1, 3))
```

We can plot the data again, but this time, we will use `new_colors` to color the data points accordingly:

```
In [12]: plot_pixels(img_data, colors=new_colors,
    ...          title="Reduce color space: 16 colors")
```

The result is the recoloring of the original pixels, where each pixel is assigned the color of its closest cluster center:

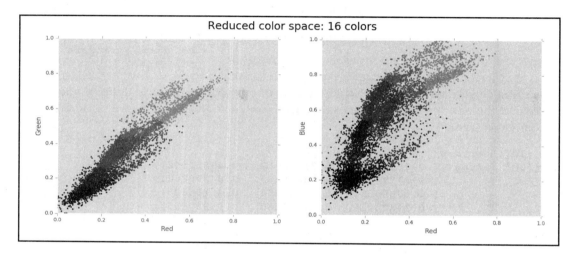

Reduced colors

In order to see the effect of this recoloring, we have to plot `new_colors` as an image. In order to get from the image to the data matrix, we flattened the earlier image. Now we need to do the inverse to get back to the image, which is to reshape `new_colors` according to the shape of the Lena image:

```
In [13]: lena_recolored = new_colors.reshape(lena.shape)
```

Then we can visualize the recolored Lena image like any other image:

```
In [14]: plt.figure(figsize=(10, 6))
    ...     plt.imshow(cv2.cvtColor(lena_recolored, cv2.COLOR_BGR2RGB));
    ...     plt.title('16-color image')
```

The result looks like this:

16-color version of Lena

Pretty cool, right?

Although some detail is arguably lost, overall, the preceding image is still clearly recognizable. This is pretty remarkable, considering that we just compressed the image by a factor of around 1 million!

You can repeat this procedure for any desired number of colors.

 Another way to reduce the color palette of images involves the use of **bilateral filters**. The resulting images often look like cartoon versions of the original image. You can find an example of this in the book *OpenCV with Python Blueprints, M. Beyeler Packt Publishing*.

Another potential application of *k*-means is something you might not expect: to use it for image classification.

Classifying handwritten digits using k-means

Although the last application was a pretty creative use of *k*-means, we can do better still. We have previously discussed *k*-means in the context of unsupervised learning, where we tried to discover some hidden structure in the data.

However, doesn't the same concept apply to most classification tasks? Let's say our task was to classify handwritten digits. Don't most zeros look similar, if not the same? And don't all zeros look categorically different from all possible ones? Isn't this exactly the kind of "hidden structure" we set out to discover with unsupervised learning? Doesn't this mean we could use clustering for classification as well?

Let's find out together. In this section, we will attempt to use *k*-means to try and classify handwritten digits. In other words, we will try to identify similar digits without using the original label information.

Loading the dataset

From the earlier chapters, you might recall that scikit-learn provides a whole range of handwritten digits via its `load_digits` utility function. The dataset consists of 1,797 samples with 64 features each, where each of the features has the brightness of one pixel in an *8 x 8* image:

```
In [1]: from sklearn.datasets import load_digits
   ...: digits = load_digits()
   ...: digits.data.shape
Out[1]: (1797, 64)
```

Running k-means

Setting up *k*-means works exactly the same as in the previous examples. We tell the algorithm to perform at most 10 iterations and stop the process if our prediction of the cluster centers does not improve within a distance of `1.0`:

```
In [2]: import cv2
   ...: criteria = (cv2.TERM_CRITERIA_EPS + cv2.TERM_CRITERIA_MAX_ITER,
   ...:             10, 1.0)
   ...: flags = cv2.KMEANS_RANDOM_CENTERS
```

Then we apply *k*-means to the data as we did before. Since there are 10 different digits (*0-9*), we tell the algorithm to look for 10 distinct clusters:

```
In [3]: import numpy as np
   ...  digits.data = digits.data.astype(np.float32)
   ...  compactness, clusters, centers = cv2.kmeans(digits.data, 10, None,
   ...                                              criteria, 10, flags)
```

And done!

Similar to the *N x 3* matrix that represented different RGB colors, this time, the centers array consists of *N x 8 x 8* center images, where *N* is the number of clusters. Therefore, if we want to plot the centers, we have to reshape the `centers` matrix back into *8 x 8* images:

```
In [4]: import matplotlib.pyplot as plt
   ...  plt.style.use('ggplot')
   ...  %matplotlib inline
   ...  fig, ax = plt.subplots(2, 5, figsize=(8, 3))
   ...  centers = centers.reshape(10, 8, 8)
   ...  for axi, center in zip(ax.flat, centers):
   ...      axi.set(xticks=[], yticks=[])
   ...      axi.imshow(center, interpolation='nearest', cmap=plt.cm.binary)
```

The output looks like this:

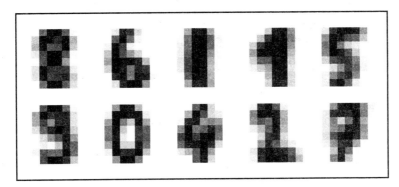

Cluster centers recovered by *k*-means

Look familiar?

Remarkably, *k*-means was able to partition the digit images not just into any 10 random clusters, but into the digits 0-9! In order to find out which images were grouped into which clusters, we need to generate a `labels` vector as we know it from supervised learning problems:

```
In [5]: from scipy.stats import mode
   ...: labels = np.zeros_like(clusters.ravel())
   ...: for i in range(10):
   ...:     mask = (clusters.ravel() == i)
   ...:     labels[mask] = mode(digits.target[mask])[0]
```

Then we can calculate the performance of the algorithm using scikit-learn's `accuracy_score` metric:

```
In [6]: from sklearn.metrics import accuracy_score
   ...: accuracy_score(digits.target, labels)
Out[6]: 0.78464106844741233
```

Remarkably, k-means achieved 78.4% accuracy without knowing the first thing about the labels of the original images!

We can gain more insights about what went wrong and how by looking at the **confusion matrix**. The confusion matrix is a 2D matrix C, where every element $C_{i,j}$ is equal to the number of observations known to be in group (or cluster) i, but predicted to be in group j. Thus, all elements on the diagonal of the matrix represent data points that have been correctly classified (that is, known to be in group i and predicted to be in group i). Off-diagonal elements show misclassifications.

In scikit-learn, creating a confusion matrix is essentially a one-liner:

```
In [7]: from sklearn.metrics import confusion_matrix
   ...: confusion_matrix(digits.target, labels)
Out[7]: array([[177,   0,   0,   0,   1,   0,   0,   0,   0,   0],
               [  0, 154,  25,   0,   0,   1,   2,   0,   0,   0],
               [  1,   3, 147,  11,   0,   0,   0,   3,  12,   0],
               [  0,   1,   2, 159,   0,   2,   0,   9,  10,   0],
               [  0,  12,   0,   0, 162,   0,   0,   5,   2,   0],
               [  0,   0,   0,  40,   2, 138,   2,   0,   0,   0],
               [  1,   2,   0,   0,   0,   0, 177,   0,   1,   0],
               [  0,  14,   0,   0,   0,   0,   0, 164,   1,   0],
               [  0,  23,   3,   8,   0,   5,   1,   2, 132,   0],
               [  0,  21,   0, 145,   0,   5,   0,   8,   1,   0]])
```

The confusion matrix tells us that k-means did a pretty good job at classifying data points from the first nine classes; however, it confused all nines to be (mostly) threes. Still, this result is pretty solid, given that the algorithm had no target labels to be trained on.

Organizing clusters as a hierarchical tree

An alternative to *k*-means is **hierarchical clustering**. One advantage of hierarchical clustering is that it allows us to organize the different clusters in a hierarchy (also known as a **dendrogram**), which can make it easier to interpret the results. Another useful advantage is that we do not need to specify the number of clusters upfront.

Understanding hierarchical clustering

There are two approaches to hierarchical clustering:

- In **agglomerative hierarchical clustering**, we start with each data point potentially being its own cluster, and we subsequently merge the closest pair of clusters until only one cluster remains.
- In **divisive hierarchical clustering**, it's the other way around: We start by assigning all the data points to one and the same cluster, and we subsequently split the cluster into smaller clusters until each cluster only contains one sample.

Of course, we can specify the number of desired clusters if we wish to. In the following figure, we asked the algorithm to find a total of three clusters:

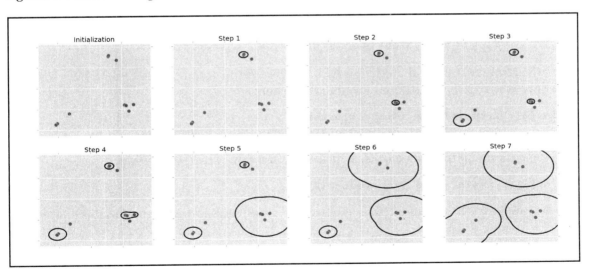

Step-by-step example of agglomerative hierarchical clustering

In step 1, the algorithm put the two closest points into their own cluster (middle, top). In step 2, two points on the lower right turned out to be the closest pair across all possible pairs in the dataset, so they were merged into their own cluster. This process continued until all data points were assigned to one of the three clusters (step 7), at which point the algorithm terminated.

 If you are using OpenCV's C++ API, you might also want to check out **Fast Approximate Nearest Neighbor (FLANN)**-based hierarchical clustering using *k*-means in `cv::flann::hierarchicalClustering`.

Let's dig deeper into implementing the agglomerative hierarchical clustering.

Implementing agglomerative hierarchical clustering

Although OpenCV does not provide an implementation of agglomerative hierarchical clustering, it is a popular algorithm that should, by all means, belong to our machine learning repertoire.

We start out by generating 10 random data points, just like in the previous figure:

```
In [1]: from sklearn.datasets import make_blobs
   ...: X, y = make_blobs(random_state=100, n_samples=10)
```

Using the familiar statistical modeling API, we import the `AgglomerativeClustering` algorithm and specify the desired number of clusters:

```
In [2]: from sklearn import cluster
   ...: agg = cluster.AgglomerativeClustering(n_clusters=3)
```

Fitting the model to the data works, as usual, via the `fit_predict` method:

```
In [3]: labels = agg.fit_predict(X)
```

We can generate a scatter plot where every data point is colored according to the predicted label:

```
In [4]: import matplotlib.pyplot as plt
   ...: %matplotlib inline
   ...: plt.style.use('ggplot')
   ...: plt.scatter(X[:, 0], X[:, 1], c=labels, s=100)
```

The resulting clustering is equivalent to the following figure:

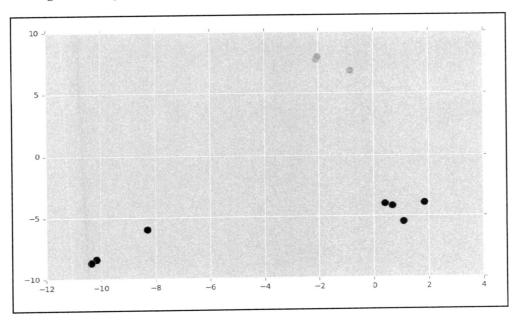

Clustering result of agglomerative hierarchical clustering

That's it! This marks the end of another wonderful adventure.

Summary

In this chapter, we talked about a number of unsupervised learning algorithms, including *k*-means, spherical clustering, and agglomerative hierarchical clustering. We saw that *k*-means is just a specific application of the more general expectation-maximization algorithm, and we discussed its potential limitations.

Furthermore, we applied *k*-means to two specific applications, which were to reduce the color palette of images and to classify handwritten digits.

In the next chapter, we will move back into the world of supervised learning and talk about some of the most powerful current machine learning algorithms: **neural networks** and **deep learning**.

9
Using Deep Learning to Classify Handwritten Digits

Let's now return to supervised learning and discuss a family of algorithms known as **artificial neural networks**. Early studies of neural networks go back to the 1940s when Warren McCulloch and Walter Pitts first described how biological nerve cells (or **neurons**) in the brain might work. More recently, artificial neural networks have seen a revival under the buzzword **deep learning**, which powers state-of-the-art technologies, such as Google's DeepMind and Facebook's DeepFace algorithms.

In this chapter, we want to wrap our heads around some simple versions of artificial neural nets, such as the **McCulloch-Pitts neuron**, the **perceptron**, and the **multilayer perceptron**. Once we have familiarized ourselves with the basics, we will be ready to implement a more sophisticated deep neural net in order to classify handwritten digits from the popular **MNIST database** (short for **Mixed National Institute of Standards and Technology database**). For this, we will be making use of **Keras**, a high-level neural network library, which is also frequently used by researchers and tech companies.

Along the way, we want to get answers to the following questions:

- How do I implement perceptrons and multilayer perceptrons in OpenCV?
- What is the difference between stochastic and batch gradient descent, and how does it fit in with backpropagation?
- How do I know what size my neural net should be?
- How can I use Keras to build sophisticated deep neural networks?

Excited? Then let's go!

Understanding the McCulloch-Pitts neuron

In 1943, Warren McCulloch and Walter Pitts published a mathematical description of neurons as they were believed to operate in the brain. A neuron receives input from other neurons through connections on its **dendritic tree**, which are integrated to produce an output at the cell body (or **soma**). The output is then communicated to other neurons via a long wire (or **axon**), which eventually branches out to make one or more connections (at **axon terminals**) on the dendritic tree of other neurons. An example neuron is shown in the following figure:

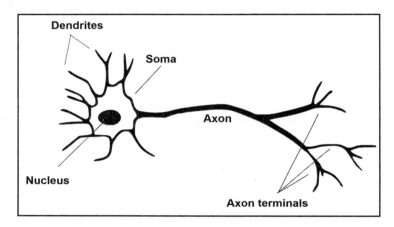

Schematic of a neuron (nerve cell). Adapted from a figure by Looxix at French Wikipedia (CC BY-SA 3.0)

McCulloch and Pitts described the inner workings of such a neuron as a simple logic gate that would be either on or off, depending on the input it receives on its dendritic tree. Specifically, the neuron would sum up all of its inputs, and if the sum exceeded a certain threshold, an output signal would be generated and passed on by the axon.

 However, today we know that real neurons are much more complicated than that. Biological neurons perform intricate nonlinear mathematical operations on thousands of inputs and can change their responsiveness dynamically depending on the context, importance, or novelty of the input signal. You can think of real neurons being as complex as computers and of the human brain being as complex as the internet.

Let's consider a simple artificial neuron that receives exactly two inputs, x_0 and x_1. The job of the artificial neuron is to calculate a sum of the two inputs (usually in the form of a **weighted sum**), and if this sum exceeds a certain threshold (often zero), the neuron will be considered active and output a one; else it will be considered silent and output a minus one (or zero). In more mathematical terms, the output y of this **McCulloch-Pitts neuron** can be described as follows:

$$y = \begin{cases} 1 & \text{if } x_0 w_0 + x_1 w_1 \geq 0 \\ -1 & \text{otherwise} \end{cases}$$

In the preceding equation, w_0 and w_1 are **weight coefficients**, which, together with x_0 and x_1, make up the weighted sum. In textbooks, the two different scenarios where the output y is either +1 and -1 would often be masked by an **activation function** ϕ, which could take on two different values:

$$y = \phi(x_o w_o + x_1 w_1) = \phi(z)$$

$$\phi(z) = \begin{cases} 1 & \text{if } z \geq \theta \\ -1 & \text{otherwise} \end{cases}$$

Here, we introduce a new variable z (the so-called **net input**), which is equivalent to the weighted sum: $z = w_0 x_0 + w_1 x_1$. The weighted sum is then compared to a threshold θ to determine the value of ϕ and subsequently the value of y. Apart from that, these two equations say exactly the same thing as the preceding one.

If these equations look strangely familiar, you might be reminded of Chapter 3, *A Taste of Machine Learning*, when we were talking about linear classifiers.

And you are right, a McCulloch-Pitts neuron is essentially a linear, binary classifier!

You can think of it this way: x_0 and x_1 are the input features, w_0 and w_1 are weights to be learned, and the classification is performed by the activation function ϕ. If we do a good job of learning the weights, which we would do with the help of a suitable training set, we could classify data as positive or negative samples. In this scenario, $\phi(z)=\theta$ would act as the decision boundary.

This might all make more sense with the help of the following figure:

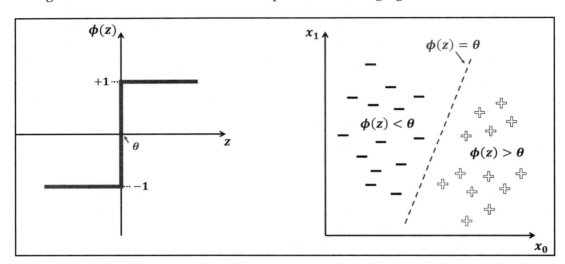

An artificial neuron's activation function (left) used to classify some example data (right)

On the left, you can see the neuron's activation function ϕ plotted against z. Remember that z is nothing more than the weighted sum of the two inputs x_0 and x_1. The rule is that as long as the weighted sum is below some threshold θ, the output of the neuron is -1; above θ, the output is +1. On the right, you can see the decision boundary denoted by $\phi(z)=\theta$, which splits the data into two regimes, $\phi(z)<\theta$ (where all data points are predicted to be negative samples) and $\phi(z)>\theta$ (where all data points are predicted to be positive samples).

The decision boundary does not need to be vertical or horizontal, it can be tilted as shown in the preceding figure. But in the case of a single McCulloch-Pitts neuron, the decision boundary will always be a straight line.

Of course, the magic lies with learning the weight coefficients w_0 and w_1 such that the decision boundary comes to lie right between all positive and all negative data points. To train a neural network, we generally need three things:

- **Training data**: It is no surprise to learn that we need some data samples with which the effectiveness of our classifier can be verified.
- **Cost function** (also known as **loss function**): A cost function provides a measure of how good the current weight coefficients are. There is a wide range of cost functions available, which we will talk about towards the end of this chapter. One solution is to count the number of misclassifications. Another one is to calculate the **sum of squared errors**.

- **Learning rule**: A learning rule specifies mathematically how we have to update the weight coefficients from one iteration to the next. This learning rule usually depends on the error (measured by the cost function) we observed on the training data.

This is where the work of renowned researcher Frank Rosenblatt comes in.

Understanding the perceptron

In the 1950s, American psychologist and artificial intelligence researcher Frank Rosenblatt invented an algorithm that would automatically learn the optimal weight coefficients w_0 and w_1 needed to perform an accurate binary classification: the **perceptron learning rule**.

Rosenblatt's original perceptron algorithm can be summed up as follows:

1. Initialize the weights to zero or some small random numbers.
2. For each training sample s_i, perform the following steps:
    1. Compute the predicted target value $\hat{y}_i$.
    2. Compare $\hat{y}_i$ to the ground truth y_i, and update the weights accordingly:
        - If the two are the same (correct prediction), skip ahead.
        - If the two are different (wrong prediction), push the weight coefficients w_0 and w_1 towards the positive or negative target class respectively.

Let's have a closer look at the last step, which is the **weight update rule**. The goal of the weight coefficients is to take on values that allow for successful classification of the data. Since we initialize the weights either to zero or small random numbers, chances that we get 100% accuracy from the outset are incredibly low. We, therefore, need to make small changes to the weights such that our classification accuracy improves.

In the two-input case, this means that we have to update w_0 by some small change Δw_0 and w_1 by some small change Δw_1:

$$w_0 = w_0 + \Delta w_0$$
$$w_1 = w_1 + \Delta w_1$$

Rosenblatt reasoned that we should update the weights in a way that makes classification more accurate. We thus need to compare the perceptron's predicted output $\hat{y}_i$ for every data point i to the ground truth y_i. If the two values are the same, the prediction was correct, meaning the perceptron is already doing a great job and we shouldn't be changing anything. However, if the two values are different, we should push the weights closer to the positive or negative class respectively. This can be summed up with the following equations:

$$\Delta w_0 = \eta \, (y_i - \hat{y}_i) \, x_0$$
$$\Delta w_1 = \eta \, (y_i - \hat{y}_i) \, x_1$$

Here, the parameter η denotes the learning rate (usually a constant between 0 and 1). Usually, η is chosen sufficiently small so that with every data sample s_i, we make only a **small step** toward the desired solution.

To get a better intuition of this update rule, let's assume that for a particular sample s_i, we predicted $\hat{y}_i = -1$, but the ground truth would have been $y_i = +1$. If we set $\eta = 1$, then we have $\Delta w_0 = (1 - (-1)) \, x_0 = 2 \, x_0$. In other words, the update rule would want to increase the weight value w_0 by $2 \, x_0$. A stronger weight value would make it more likely for the weighted sum to be above the threshold next time, thus hopefully classifying the sample as $\hat{y}_i = -1$ next time around.

On the other hand, if $\hat{y}_i$ is the same as y_i, then $(\hat{y}_i - y_i)$ would cancel out, and $\Delta w_0 = 0$.

The perceptron learning rule is **guaranteed to converge** on the optimal solution only if the data is linearly separable and the learning rate is sufficiently small.

It is straightforward—in a mathematical sense—to extend the perceptron learning rule to more than two inputs x_0 and x_1 by extending the number of terms in the weighted sum:

$$z = x_o w_o + x_1 w_1 + \cdots + x_m w_m = \sum_{i=1}^{m} x_i w_i$$

$$\phi(z) = \begin{cases} 1 & \text{if } z \geq \theta \\ -1 & \text{otherwise} \end{cases}$$

It is customary to make one of the weight coefficients not depend on any input features (usually w_0, so $x_0 = 1$) so that it can act as a scalar (or **bias term**) in the weighted sum.

To make the perceptron algorithm look more like an artificial neuron, we can illustrate it as follows:

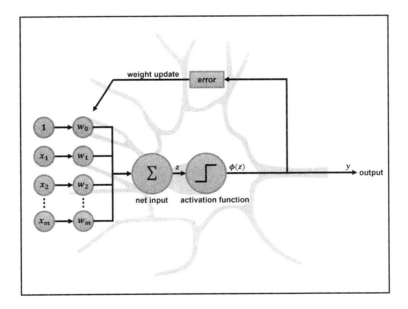

Illustrating the perceptron algorithm as an artificial neuron

Here, the input features and weight coefficients are interpreted as part of the dendritic tree, where the neuron gets all its inputs from. These inputs are then summed up and passed through the activation function, which happens at the cell body. The output is then sent along the axon to the next cell. In the perceptron learning rule, we also made use of the neuron's output to calculate the error, which, in turn, was used to update the weight coefficients.

One distinction between the 1950s version of the perceptron and modern perceptrons is the use of a more sophisticated activation function ϕ; for example, the hyperbolic tangent or the sigmoid function.

Now, let's try it on some sample data!

Implementing your first perceptron

Perceptrons are easy enough to be implemented from scratch. We can mimic the typical OpenCV or scikit-learn implementation of a classifier by creating a Perceptron object. This will allow us to initialize new perceptron objects that can learn from data via a fit method and make predictions via a separate predict method.

When we initialize a new perceptron object, we want to pass a learning rate (lr, or η in the previous section) and the number of iterations after which the algorithm should terminate (n_iter):

```
In [1]: import numpy as np
In [2]: class Perceptron(object):
   ...      def __init__(self, lr=0.01, n_iter=10):
   ...          self.lr = lr
   ...          self.n_iter = n_iter
   ...
```

The fit method is where most of the work is done. This method should take as input some data samples (X) and their associated target labels (y). We will then create an array of weights (self.weights), one for each feature (X.shape[1]), initialized to zero. For convenience, we will keep the bias term (self.bias) separate from the weight vector and initialize it to zero as well:

```
   ...      def fit(self, X, y):
   ...          self.weights = np.zeros(X.shape[1])
   ...          self.bias = 0.0
```

The predict method should take in a number of data samples (X) and for each of them, return a target label, either +1 or -1. In order to perform this classification, we need to implement $\phi(z)>\theta$. Here we will choose $\theta = 0$, and the weighted sum can be computed with NumPy's dot product:

```
   ...      def predict(self, X):
   ...          return np.where(np.dot(X, self.weights) + self.bias >= 0.0,
   ...                          1, -1)
```

Then we will calculate the Δw terms for every data sample (xi, yi) in the dataset and repeat this step for a number of iterations (self.n_iter). For this, we need to compare the ground-truth label (yi) to the predicted label (aforementioned self.predict(xi)). The resulting delta term will be used to update both the weights and the bias term:

```
...             for _ in range(self.n_iter):
...                 for xi, yi in zip(X, y):
...                     delta = self.lr * (yi - self.predict(xi))
...                     self.weights += delta * xi
...                     self.bias += delta
```

That's it!

Generating a toy dataset

To test our perceptron classifier, we need to create some mock data. Let's keep things simple for now and generate 100 data samples (n_samples) belonging to one of two blobs (centers), again relying on scikit-learn's make_blobs function:

```
In [3]: from sklearn.datasets.samples_generator import make_blobs
...     X, y = make_blobs(n_samples=100, centers=2,
...                       cluster_std=2.2, random_state=42)
```

One thing to keep in mind is that our perceptron classifier expects target labels to be either +1 or -1, whereas make_blobs returns 0 and 1. An easy way to adjust the labels is with the following equation:

```
In [4]: y = 2 * y - 1
```

Let's have a look at the data:

```
In [5]: import matplotlib.pyplot as plt
...     plt.style.use('ggplot')
...     %matplotlib inline
...     plt.scatter(X[:, 0], X[:, 1], s=100, c=y);
...     plt.xlabel('x1')
...     plt.ylabel('x2')
```

This will produce the following plot:

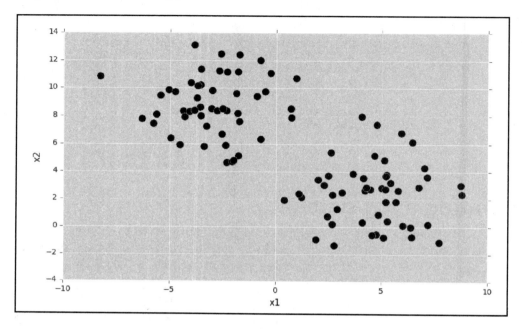

Example dataset for the perceptron classifier

What do you think, will our perceptron classifier have an easy time finding a decision boundary that separates these two blobs?

Chances are it will. We mentioned earlier that a perceptron is a linear classifier. This means that as long as you can draw a straight line in the preceding plot to separate the two blobs, there exists a linear decision boundary that the perceptron should be able to find, that is, if we implemented everything correctly. Let's find out.

Fitting the perceptron to data

We can instantiate our perceptron object similar to other classifiers we encountered with OpenCV:

```
In [6]: p = Perceptron(lr=0.1, n_iter=10)
```

Here, we chose a learning rate of 0.1 and told the perceptron to terminate after 10 iterations. These values are chosen rather arbitrarily at this point, although we will come back to them in a little while.

Choosing an appropriate learning rate is critical, but it's not always clear what the most appropriate choice is. The learning rate determines how fast or slow we move towards the optimal weight coefficients. If the learning rate is too large, we might accidentally skip the optimal solution. If it is too small, we will need a large number of iterations to converge to the best values.

Once the perceptron is set up, we can call the `fit` method to optimize the weight coefficients:

```
In [7]: p.fit(X, y)
```

Did it work? Let's have a look at the learned weight values:

```
In [8]: p.weights
Out[8]: array([ 2.20091094, -0.4798926 ])
```

And don't forget to have a peek at the bias term:

```
In [9]: p.bias
Out[9]: 0.20000000000000001
```

If we plug these values into our equation for ϕ, it becomes clear that the perceptron learned a decision boundary of the form $2.2\,x_1 - 0.48\,x_2 + 0.2 \geq 0$.

Evaluating the perceptron classifier

In order to find out how good our perceptron performs, we can calculate the accuracy score on all data samples:

```
In [10]: from sklearn.metrics import accuracy_score
    ...: accuracy_score(p.predict(X), y)
Out[10]: 1.0
```

Perfect score!

Let's have a look at the decision landscape by bringing back our `plot_decision_boundary` from the earlier chapters:

```
In [10]: def plot_decision_boundary(classifier, X_test, y_test):
    ...:     # create a mesh to plot in
    ...:     h = 0.02 # step size in mesh
    ...:     x_min, x_max = X_test[:, 0].min() - 1, X_test[:, 0].max() + 1
    ...:     y_min, y_max = X_test[:, 1].min() - 1, X_test[:, 1].max() + 1
    ...:     xx, yy = np.meshgrid(np.arange(x_min, x_max, h),
```

```
...
...                               np.arange(y_min, y_max, h))
...        X_hypo = np.c_[xx.ravel().astype(np.float32),
...                       yy.ravel().astype(np.float32)]
...        zz = classifier.predict(X_hypo)
...        zz = zz.reshape(xx.shape)
...
...        plt.contourf(xx, yy, zz, cmap=plt.cm.coolwarm, alpha=0.8)
...        plt.scatter(X_test[:, 0], X_test[:, 1], c=y_test, s=200)
```

We can plot the decision landscape by passing the perceptron object (p), the data (X), and the corresponding target labels (y):

```
In [11]: plot_decision_boundary(p, X, y)
```

This will produce the following plot:

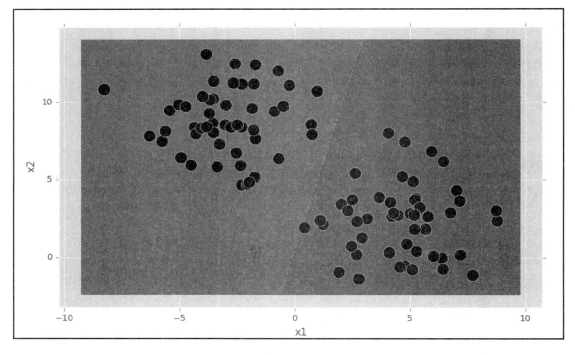

Linear decision boundary that separates the two classes

Of course, this problem was rather simple, even for a simple linear classifier. Even more so, you might have noticed that we didn't split the data into training and test sets. If we had data that wasn't linearly separable, the story might be a little different.

Applying the perceptron to data that is not linearly separable

Since the perceptron is a linear classifier, you can imagine that it would have trouble trying to classify data that is not linearly separable. We can test this by increasing the spread (cluster_std) of the two blobs in our toy dataset so that the two blobs start overlapping:

```
In [12]: X, y = make_blobs(n_samples=100, centers=2,
   ...        cluster_std=5.2, random_state=42)
   ...   y = 2 * y - 1
```

We can plot the dataset again using Matplotlib's scatter function:

```
In [13]: plt.scatter(X[:, 0], X[:, 1], s=100, c=y);
   ...    plt.xlabel('x1')
   ...    plt.ylabel('x2')
```

As is evident in the following figure, this data is no longer linearly separable because there is no straight line that perfectly separates the two blobs:

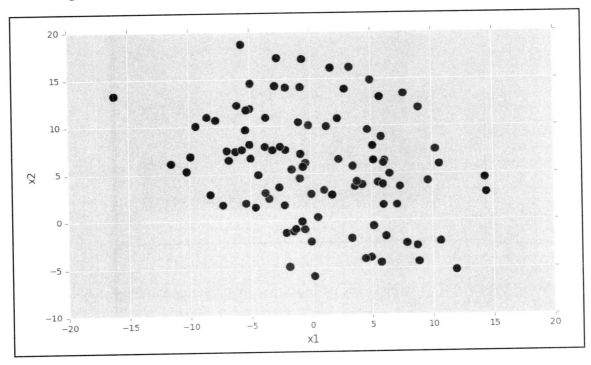

Example data that is not linearly separable

So what would happen if we applied the perceptron classifier to this dataset?

We can find an answer to this question by repeating the preceding steps:

```
In [14]: p = Perceptron(lr=0.1, n_iter=10)
   ...:  p.fit(X, y)
```

Then we find an accuracy score of 81%:

```
In [15]: accuracy_score(p.predict(X), y)
Out[15]: 0.81000000000000005
```

In order to find out which data points were misclassified, we can again visualize the decision landscape using our helper function:

```
In [16]: plot_decision_boundary(p, X, y)
   ...:  plt.xlabel('x1')
   ...:  plt.ylabel('x2')
```

The following figure makes the limitations of the perceptron classifier evident. Being a linear classifier, it tried to separate the data using a straight line but ultimately failed:

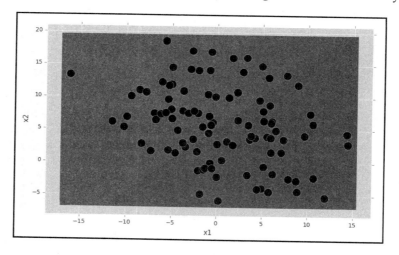

The perceptron's linear decision boundary can no longer separate the data

Fortunately, there are ways to make the perceptron more powerful and ultimately create nonlinear decision boundaries.

Understanding multilayer perceptrons

In order to create nonlinear decision boundaries, we can combine multiple perceptrons to form a larger network. This is also known as a **multilayer perceptron** (**MLP**). MLPs usually consist of at least three layers, where the first layer has a node (or neuron) for every input feature of the dataset, and the last layer has a node for every class label. The layer in between is called the **hidden layer**. An example of this **feedforward neural network** architecture is shown in the following figure:

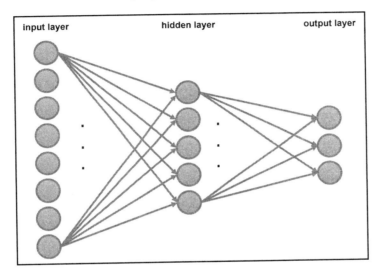

Example of a three-layer perceptron

In this network, every circle is an artificial neuron (or essentially a perceptron), and the output of one artificial neuron might serve as input to the next artificial neuron, much like how real biological neurons are wired up in the brain. By placing perceptrons side by side, we get a single one-layer neural network. Analogously, by stacking one one-layer neural network upon the other, we get a multilayer neural network—an MLP.

 A neural network that contains loops and/or backward connections is called a **recurrent neural network**.

One remarkable property of MLPs is that they can represent any mathematical function if you make the network big enough. This is also known as the **universal approximation property**. For example, a one hidden-layer MLP (such as the one shown in the preceding figure) is able to exactly represent:

- Any boolean function (such as AND, OR, NOT, NAND, and so on)
- Any bounded continuous function (such as the sine or cosine functions)

Even better, if you add another layer, an MLP can approximate any arbitrary function. You can see why these neural networks are so powerful—if you just give them enough neurons and enough layers, they can basically learn any input-output function!

Understanding gradient descent

When we talked about the perceptron earlier in this chapter, we identified three essential ingredients needed for training: training data, a cost function, and a learning rule. Whereas the learning rule worked great for a single perceptron, unfortunately, it did not generalize to MLPs, so people had to come up with a more general rule.

If you think about how we measure the success of a classifier, we usually do so with the help of a cost function. A typical example is the number of misclassifications of the network or the mean squared error. This function (also known as a **loss function**) usually depends on the parameters we are trying to tweak. In neural networks, these parameters are the weight coefficients. Let's assume a simple neural net has a single weight to tweak, w. Then we can visualize the cost as a function of the weight:

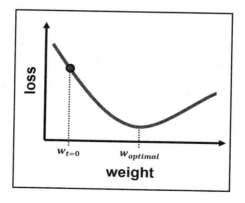

Example loss function

At the beginning of training, at time zero, we may start out way on the left of this graph ($w_{t=0}$). But from the graph, we know that there would be a better value for w, namely $w_{optimal}$, which would minimize the cost function. Smallest cost means lowest error, so it should be our highest goal to reach $w_{optimal}$ through learning.

This is exactly what **gradient descent** does. You can think of the gradient as a vector that points up the hill. In gradient descent, we are trying to walk opposite of the gradient, effectively walking down the hill, from the peaks down to the valley:

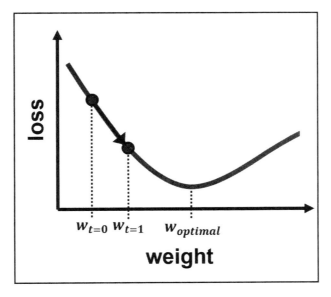

Gradient descent

Once you reach the valley, the gradient goes to zero, and that completes the training.

There are several ways to reach the valley—we could approach from the left, or we could approach from the right. The starting point of our descent is determined by the **initial weight values**. Furthermore, we have to be careful not to take too large a step, otherwise we might miss the valley:

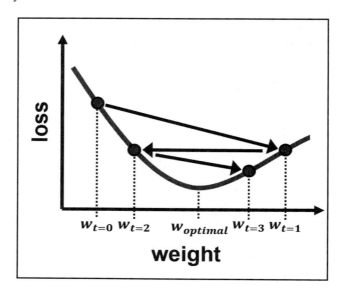

Consequences of choosing too large a step size in gradient descent

Hence, in **stochastic gradient descent** (sometimes also called **iterative** or **on-line gradient descent**), the goal is to take small steps but to take them as often as possible. The effective step size is determined by the **learning rate** of the algorithm. Specifically, we would perform the following procedure over and over:

1. Present a small number of training samples to the network (called the **batch size**).
2. On this small batch of data, calculate the gradient of the cost function.
3. Update the weight coefficients by taking a small step in the opposite direction of the gradient, towards the valley.
4. Repeat steps 1-3 until the weight cost no longer goes down. This is an indication that we have reached the valley.

A compromise between batch gradient descent and stochastic gradient descent is the so-called **mini-batch learning**. Mini-batch learning can be understood as applying batch gradient descent to smaller subsets of the training data, for example, 50 samples at a time.

Can you think of an example where this procedure might fail?

One scenario that comes to mind is where the cost function has multiple valleys, some deeper than others, as shown in the following figure:

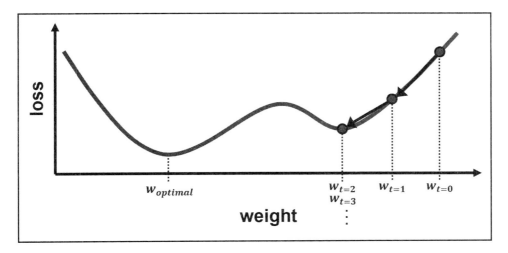

Getting stuck in a local minimum

If we start on the left, we should arrive at the same valley as before—no problem. But, if our starting point is all the way to the right, we might encounter another valley on the way. Gradient descent will lead us straight down to the valley, but it will not have any means to climb out of it.

 This is also known as **getting stuck in a local minimum**. Researchers have come up with different ways to try and avoid this issue, one of them being to add noise to the process.

There is one piece left in the puzzle. Given our current weight coefficients, how do we know the slope of the cost function?

Training multi-layer perceptrons with backpropagation

This is where **backpropagation** comes in, which is an algorithm for estimating the gradient of the cost function in neural networks. Some might say that it is basically a fancy word for the chain rule, which is a means to calculate the **partial derivative** of functions that depend on more than one variable. Nonetheless, it is a method that helped bring the field of artificial neural networks back to life, so we should be thankful for that.

Understanding backpropagation involves quite a bit of calculus, so I will only give you an intuition here.

Let's remind ourselves that the cost function, and therefore its gradient, depends on the difference between the true output (y_i) and the current output ($\hat{y}_i$) for every data sample i. If we choose to define the cost function as the mean squared error, the corresponding equation would look like this:

$$E = \frac{1}{2} \sum_i (y_i - \hat{y}_i)^2$$

Here, the sum is over all data samples i. Of course, the output of a neuron ($\hat{y}$) depends on its input (z), which, in turn, depends on input features (x_j) and weights (w_j):

$$\hat{y} = \phi(z)$$

$$z = \sum_j x_j w_j$$

Therefore, in order to calculate the slope of the cost function as shown in the preceding figures, we have to calculate the derivative of E with respect to some weight, w_j. This calculation can involve quite a bit of mathematics because E depends on $\hat{y}$, which depends on z, which depends on w_j. However, thanks to the chain rule, we know that we can compartmentalize this calculation in the following way:

$$\frac{\partial E}{\partial w_j} = \frac{\partial E}{\partial \hat{y}} \frac{\partial \hat{y}}{\partial z} \frac{\partial z}{\partial w_j}$$

Here, the term of the left-hand side is the partial derivative of E with respect to w_j, which is what we are trying to calculate. If we had more neurons in the network, the equation would not end at w_j but might include some additional terms describing how the error flows through different hidden layers. This is also where backpropagation gets its name from; similar to the activity flowing from the input layer via the hidden layers to the output layer, the error gradient flows backwards through the network. In other words, the error gradient in the hidden layer depends on the error gradient in the output layer, and so on. You can see it is as if the error **back-propagated** through the network.

Well, this is still kind of complicated, you might say. The good news is, however, that the only term to depend on w_j in the preceding equation is the rightmost term, making the math at least a little easier. But you are right, this is still kind of complicated, and there's no reason for us to dig deeper here.

For more information on backpropagation, I can warmly recommend the following book: Ian Goodfellow, Yoshua Bengio, and Aaron Courville (2016). *Deep Learning (Adaptive Computation and Machine Learning series)*. MIT Press, ISBN 978-026203561-3.

After all, what we are really interested in is how all this stuff works in practice.

Implementing a multilayer perceptron in OpenCV

Implementing an MLP in OpenCV uses the same syntax that we have seen at least a dozen times before. In order to see how an MLP compares to a single perceptron, we will operate on the same toy data as before:

```
In [1]: from sklearn.datasets.samples_generator import make_blobs
   ...      X_raw, y_raw = make_blobs(n_samples=100, centers=2,
   ...                                cluster_std=5.2, random_state=42)
```

Preprocessing the data

However, since we are working with OpenCV, this time, we want to make sure the input matrix is made up of 32-bit floating point numbers, otherwise the code will break:

```
In [2]: import numpy as np
   ...    X = X_raw.astype(np.float32)
```

Furthermore, we need to think back to Chapter 4, *Representing Data and Engineering and Features*, and remember how to represent categorical variables. We need to find a way to represent target labels, not as integers but with a one-hot encoding. The easiest way to achieve this is by using scikit-learn's preprocessing module:

```
In [3]: from sklearn.preprocessing import OneHotEncoder
   ...    enc = OneHotEncoder(sparse=False, dtype=np.float32)
   ...    y = enc.fit_transform(y_raw.reshape(-1, 1))
```

Creating an MLP classifier in OpenCV

The syntax to create an MLP in OpenCV is the same as for all the other classifiers:

```
In [4]: import cv2
   ...    mlp = cv2.ml.ANN_MLP_create()
```

However, now we need to specify how many layers we want in the network and how many neurons there are per layer. We do this with a list of integers, which specify the number of neurons in each layer. Since the data matrix X has two features, the first layer should also have two neurons in it (n_input). Since the output has two different values, the last layer should also have two neurons in it (n_output). In between these two layers, we can put as many hidden layers with as many neurons as we want. Let's choose a single hidden layer with an arbitrary number of eight neurons in it (n_hidden):

```
In [5]: n_input = 2
   ...    n_hidden = 8
   ...    n_output = 2
   ...    mlp.setLayerSizes(np.array([n_input, n_hidden, n_output]))
```

Customizing the MLP classifier

Before we move on to training the classifier, we can customize the MLP classifier via a number of optional settings:

- `mlp.setActivationFunction`: This defines the activation function to be used for every neuron in the network
- `mlp.setTrainMethod`: This defines a suitable training method
- `mlp.setTermCriteria`: This sets the termination criteria of the training phase

Whereas our home-brewed perceptron classifier used a linear activation function, OpenCV provides two additional options:

- `cv2.ml.ANN_MLP_IDENTITY`: This is the linear activation function, $f(x) = x$.
- `cv2.ml.ANN_MLP_SIGMOID_SYM`: This is the symmetrical sigmoid function (also known as **hyperbolic tangent**), $f(x) = \beta (1 - \exp(-\alpha x)) / (1 + \exp(-\alpha x))$. Whereas α controls the slope of the function, β defines the upper and lower bounds of the output.
- `cv2.ml.ANN_GAUSSIAN`: This is the Gaussian function (also known as the **bell curve**), $f(x) = \beta \exp(-\alpha x^2)$. Whereas α controls the slope of the function, β defines the upper bound of the output.

> Careful, OpenCV is being really confusing here! What they call a symmetrical sigmoid function is in reality the hyperbolic tangent. In other software, a sigmoid function usually has responses in the range [0, 1], whereas the hyperbolic tangent has responses in the range [-1, 1]. Even worse, if you use the symmetrical sigmoid function with its default parameters, the output will range in [-1.7159, 1.7159]!

In this example, we will use a proper sigmoid function that squashes the input values into the range [0, 1]. We do this by choosing $\alpha = 2.5$ and $\beta = 1.0$:

```
In [6]: mlp.setActivationFunction(cv2.ml.ANN_MLP_SIGMOID_SYM, 2.5, 1.0)
```

If you are curious what this activation function looks like, we can take a short excursion with Matplotlib:

```
In [7]: import matplotlib.pyplot as plt
   ...:     %matplotlib inline
   ...:     plt.style.use('ggplot')
```

In order to see what the activation function looks like, we can create a NumPy array that densely samples x values in the range [-1, 1], and then calculate the corresponding y values using the preceding mathematical expression:

```
In [8]: alpha = 2.5
   ...  beta = 1.0
   ...  x_sig = np.linspace(-1.0, 1.0, 100)
   ...  y_sig = beta * (1.0 - np.exp(-alpha * x_sig))
   ...  y_sig /= (1 + np.exp(-alpha * x_sig))
   ...  plt.plot(x_sig, y_sig, linewidth=3);
   ...  plt.xlabel('x')
   ...  plt.ylabel('y')
```

As you can see in the following figure, this will create a nice squashing function whose output values will lie in the range [-1, 1]:

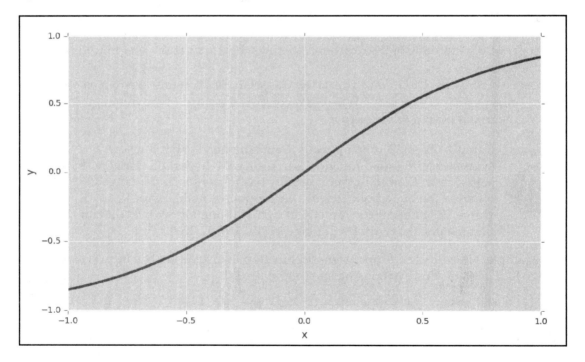

Example of the symmetrical sigmoid function for $\alpha = 2.5$ and $\beta = 1.0$

As mentioned in the preceding part, a training method can be set via `mlp.setTrainMethod`. The following methods are available:

- `cv2.ml.ANN_MLP_BACKPROP`: This is the backpropagation algorithm we talked about previously. You can set additional scaling factors via `mlp.setBackpropMomentumScale` and `mlp.setBackpropWeightScale`.
- `cv2.ml.ANN_MLP_RPROP`: This is the Rprop algorithm, which is short for resilient backpropagation. We won't have time to discuss this algorithm, but you can set additional parameters of this algorithm via `mlp.setRpropDW0`, `mlp.setRpropDWMax`, `mlp.setRpropDWMin`, `mlp.setRpropDWMinus`, and `mlp.setRpropDWPlus`.

In this example, we will choose backpropagation:

```
In [9]: mlp.setTrainMethod(cv2.ml.ANN_MLP_BACKPROP)
```

Lastly, we can specify the criteria that must be met for training to end via `mlp.setTermCriteria`. This works the same for every classifier in OpenCV and is closely tied to the underlying C++ functionality. We first tell OpenCV which criteria we are going to specify (for example, the maximum number of iterations). Then we specify the value for this criterion. All values must be delivered in a tuple.

Hence, in order to run our MLP classifier until we either reach 300 iterations or the error does not increase anymore beyond some small range of values, we would write the following:

```
In [10]: term_mode = cv2.TERM_CRITERIA_MAX_ITER + cv2.TERM_CRITERIA_EPS
    ...: term_max_iter = 300
    ...: term_eps = 0.01
    ...: mlp.setTermCriteria((term_mode, term_max_iter, term_eps))
```

Then we are ready to train the classifier!

Training and testing the MLP classifier

This is the easy part. Training the MLP classifier is the same as with all other classifiers:

```
In [11]: mlp.train(X, cv2.ml.ROW_SAMPLE, y)
Out[11]: True
```

The same goes for predicting target labels:

```
In [12]: _, y_hat = mlp.predict(X)
```

The easiest way to measure accuracy is by using scikit-learn's helper function:

```
In [13]: from sklearn.metrics import accuracy_score
   ...:  accuracy_score(y_hat.round(), y)
Out[13]: 0.83999999999999997
```

It looks like we were able to increase our performance from 81% with a single perceptron to 84% with an MLP consisting of ten hidden-layer neurons and two output neurons. In order to see what changed, we can look at the decision boundary one more time:

```
In [14]: def plot_decision_boundary(classifier, X_test, y_test):
   ...:      # create a mesh to plot in
   ...:      h = 0.02 # step size in mesh
   ...:      x_min, x_max = X_test[:, 0].min() - 1, X_test[:, 0].max() + 1
   ...:      y_min, y_max = X_test[:, 1].min() - 1, X_test[:, 1].max() + 1
   ...:      xx, yy = np.meshgrid(np.arange(x_min, x_max, h),
   ...:                           np.arange(y_min, y_max, h))
   ...:
   ...:      X_hypo = np.c_[xx.ravel().astype(np.float32),
   ...:                     yy.ravel().astype(np.float32)]
   ...:      _, zz = classifier.predict(X_hypo)
```

However, there is a problem right here, in that `zz` is now a one-hot encoded matrix. In order to transform the one-hot encoding into a number that corresponds to the class label (zero or one), we can use NumPy's `argmax` function:

```
   ...:      zz = np.argmax(zz, axis=1)
```

Then the rest stays the same:

```
   ...:      zz = zz.reshape(xx.shape)
   ...:      plt.contourf(xx, yy, zz, cmap=plt.cm.coolwarm, alpha=0.8)
   ...:      plt.scatter(X_test[:, 0], X_test[:, 1], c=y_test, s=200)
```

Then we can call the function like this:

```
In [15]: plot_decision_boundary(mlp, X, y_raw)
```

The output looks like this:

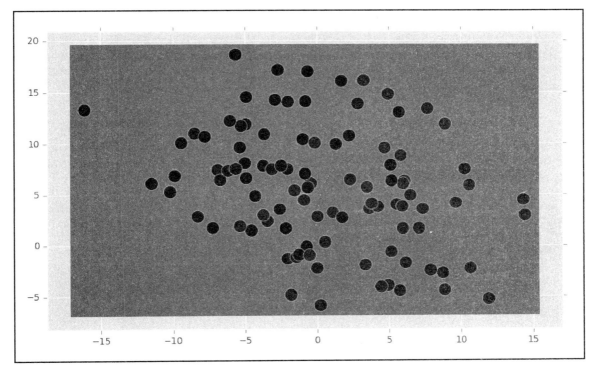

Decision boundary of an MLP with one hidden layer

And voila! The decision boundary is no longer a straight line. That being said, you might have expected a more drastic performance increase. But nobody said we have to stop here! There are at least two different things we can try from here on out:

- We can add more neurons to the hidden layer. You can do this by replacing n_hidden on line 6 with a larger value and running the code again. Generally speaking, the more neurons you put in the network, the more powerful the MLP will be.
- We can add more hidden layers. It turns out that this is where neural nets really get their power from.

Hence, this is where I should tell you about deep learning.

Getting acquainted with deep learning

Back when deep learning didn't have a fancy name yet, it was called artificial neural networks. So you already know a great deal about it! This was a respected field in itself, but after the days of Rosenblatt's perceptron, many researchers and machine learning practitioners slowly began to lose interest in the field since no one had a good solution for training a neural network with multiple layers.

Eventually, interest in neural networks was rekindled in 1986 when *David Rumelhart*, *Geoffrey Hinton*, and *Ronald Williams* were involved in the (re)discovery and popularization of the aforementioned backpropagation algorithm. However, it was not until recently that computers became powerful enough so they could actually execute the backpropagation algorithm on large-scale networks, leading to a surge in deep learning research.

You can find more information on the history and origin of deep learning in the following scientific article: Wang & Raj (2017), *On the Origin of Deep Learning*, arXiv:1702.07800.

With the current popularity of deep learning in both industry and academia, we are fortunate enough to have a whole range of open-source deep learning frameworks at our disposal:

- Google Brain's **TensorFlow** (http://www.tensorflow.org): This is a machine learning library that describes computations as dataflow graphs. To date, this is one of the most commonly used deep learning libraries. Hence, it is also evolving quickly, so you might have to check back often for software updates. TensorFlow provides a whole range of user interfaces, including Python, C++, and Java interface.
- Microsoft Research's **Cognitive Toolkit** (CNTK, https://www.microsoft.com/en-us/research/product/cognitive-toolkit): This is a deep learning framework that describes neural networks as a series of computational steps via a directed graph.
- UC Berkeley's **Caffe** (http://caffe.berkeleyvision.org): This is a pure deep learning framework written in C++, with an additional Python interface.

- University of Montreal's **Theano** (`http://deeplearning.net/software/theano`): This is a numerical computation library compiled to run efficiently on CPU and GPU architectures. Theano is more than a machine learning library; it can express any computation using a specialized computer algebra system. Hence, it is best suited for people who wish to write their machine learning algorithms from scratch.
- **Torch** (`http://www.torch.ch`): This is a scientific computing framework based on the Lua programming language. Like Theano, Torch is more than a machine learning library, but it is heavily used for deep learning by companies such as Facebook, IBM, and Yandex.

Finally, there is also **Keras**, which we will be using in the following sections. In contrast to the preceding frameworks, Keras understands itself as an interface rather than an end-to-end deep learning framework. It allows you to specify deep neural nets using an easy-to-understand API, which can then be run on backends, such as TensorFlow, CNTK, or Theano.

Getting acquainted with Keras

The core data structure of Keras is a model, which is similar to OpenCV's classifier object, except it focuses on neural networks only. The simplest type of model is the `Sequential` model, which arranges the different layers of the neural net in a linear stack, just like we did for the MLP in OpenCV:

```
In [1]: from keras.models import Sequential
   ...: model = Sequential()
Out[1]: Using TensorFlow backend.
```

Then different layers can be added to the model one by one. In Keras, layers do not just contain neurons, they also perform a function. Some core layer types include the following:

- `Dense`: This is a densely connected layer. This is exactly what we used when we designed our MLP: a layer of neurons that is connected to every neuron in the previous layer.
- `Activation`: This applies an activation function to an output. Keras provides a whole range of activation functions, including OpenCV's identify function (`linear`), the hyperbolic tangent (`tanh`), a sigmoidal squashing function (`sigmoid`), a softmax function (`softmax`), and many more.
- `Reshape`: This reshapes an output to a certain shape.

There are other layers that calculate arithmetic or geometric operations on their inputs:

- **Convolutional layers**: These layers allow you to specify a kernel with which the input layer is convolved. This allows you to perform operations such as a Sobel filter or apply a Gaussian kernel in 1D, 2D, or even 3D.
- **Pooling layers**: These layers perform a max pooling operation on their input, where the output neuron's activity is given by the maximally active input neuron.

Some other layers that are popular in deep learning are as follows:

- `Dropout`: This layer randomly sets a fraction of input units to zero at each update. This is a way to inject noise into the training process, making it more robust.
- `Embedding`: This layer encodes categorical data, similar to some functions from scikit-learn's `preprocessing` module.
- `GaussianNoise`: This layer applies additive zero-centered Gaussian noise. This is another way of injecting noise into the training process, making it more robust.

A perceptron similar to the preceding one could thus be implemented using a `Dense` layer that has two inputs and one output. Staying true to our earlier example, we will initialize the weights to zero and use the hyperbolic tangent as an activation function:

```
In [2]: from keras.layers import Dense
   ...     model.add(Dense(1, activation='tanh', input_dim=2,
   ...                     kernel_initializer='zeros'))
```

Finally, we want to specify the training method. Keras provides a number of optimizers, including the following:

- **stochastic gradient descent** (`'sgd'`): This is what we have discussed before
- **root mean square propagation** (`'RMSprop'`): This is a method in which the learning rate is adapted for each of the parameters
- **adaptive moment estimation** (`'Adam'`): This is an update to root mean square propagation and many more

In addition, Keras also provides a number of different loss functions:

- **mean squared error** (`'mean_squared_error'`): This is what was discussed before
- **hinge loss** (`'hinge'`): This is a maximum-margin classifier often used with SVM, as discussed in Chapter 6, *Detecting Pedestrians with Support Vector Machines*, and many more

You can see that there's a whole plethora of parameters to be specified and methods to choose from. To stay true to our aforementioned perceptron implementation, we will choose stochastic gradient descent as an optimizer, the mean squared error as a cost function, and accuracy as a scoring function:

```
In [3]: model.compile(optimizer='sgd',
   ...                loss='mean_squared_error',
   ...                metrics=['accuracy'])
```

In order to compare the performance of the Keras implementation to our home-brewed version, we will apply the classifier to the same dataset:

```
In [4]: from sklearn.datasets.samples_generator import make_blobs
   ...     X, y = make_blobs(n_samples=100, centers=2,
   ...                       cluster_std=2.2, random_state=42)
```

Finally, a Keras model is fit to the data with a very familiar syntax. Here, we can also choose how many iterations to train for (`epochs`), how many samples to present before we calculate the error gradient (`batch_size`), whether to shuffle the dataset (`shuffle`), and whether to output progress updates (`verbose`):

```
In [5]: model.fit(X, y, epochs=400, batch_size=100, shuffle=False,
   ...            verbose=0)
```

After the training completes, we can evaluate the classifier as follows:

```
In [6]: model.evaluate(X, y)
Out[6]: 32/100 [========>.................] - ETA: 0s
        [0.040941802412271501, 1.0]
```

Here, the first reported value is the mean squared error, whereas the second value denotes accuracy. This means that the final mean squared error was 0.04, and we had 100% accuracy. Way better than our own implementation!

You can find more information on Keras, source code documentation, and a number of tutorials at `http://keras.io`.

With these tools in hand, we are now ready to approach a real-world dataset!

Classifying handwritten digits

In the previous section, we covered a lot of the theory around neural networks, which can be a little overwhelming if you are new to this topic. In this section, we will classify handwritten digits from the popular **MNIST dataset**, which has been constructed by Yann LeCun and colleagues and serves as a popular benchmark dataset for machine learning algorithms.

We will train two different networks on it:

- a multilayer perceptron using OpenCV
- a deep neural net using Keras

Loading the MNIST dataset

The easiest way to obtain the MNIST dataset is using Keras:

```
In [1]: from keras.datasets import mnist
   ...      (X_train, y_train), (X_test, y_test) = mnist.load_data()
Out[1]: Using TensorFlow backend.
        Downloading data from
        https://s3.amazonaws.com/img-datasets/mnist.npz
```

This will download the data from the Amazon Cloud (might take a while depending on your internet connection) and automatically split the data into training and test sets.

 MNIST provides its own predefined train-test split. This way, it is easier to compare performance of different classifiers because they will all use the same data for training and the same data for testing.

This data comes in a format that we are already familiar with:

```
In [2]: X_train.shape, y_train.shape
Out[2]: ((60000, 28, 28), (60000,))
```

We should take note that the labels come as integer values between zero and nine (corresponding to the digits 0-9):

```
In [3]: import numpy as np
   ...  np.unique(y_train)
Out[3]: array([0, 1, 2, 3, 4, 5, 6, 7, 8, 9], dtype=uint8)
```

We can have a look at some example digits:

```
In [4]: import matplotlib.pyplot as plt
   ...  %matplotlib inline
In [5]: for i in range(10):
   ...      plt.subplot(2, 5, i + 1)
   ...      plt.imshow(X_train[i, :, :], cmap='gray')
   ...      plt.axis('off')
```

The digits look like this:

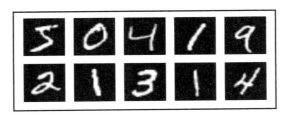

Example digits from the MNIST database

In fact, the MNIST dataset is the successor to the NIST digits dataset provided by scikit-learn that we used before (`sklearn.datasets.load_digits` (refer to Chapter 2, *Working with Data in OpenCV and Python*). Some notable differences are as follows:

- MNIST images are significantly larger (28x28 pixels) than NIST images (8x8 pixels), thus paying more attention to fine details, such as distortions and individual differences between images of the same digit
- The MNIST dataset is much larger than the NIST dataset, providing 60,000 training and 10,000 test samples (as compared to a total of 5,620 NIST images)

Preprocessing the MNIST dataset

As we learned in `Chapter 4`, *Representing Data and Engineering Features*, there are a number of preprocessing steps we might like to apply here:

- **Centering**: It is important that all the digits are centered in the image. For example, take a look at all the example images of the digit 1 in the preceding figure, which are all made of an almost-vertical strike. If the images were misaligned, the strike could lie anywhere in the image, making it hard for the neural network to find commonalities in the training samples. Fortunately, images in MNIST are already centered.
- **Scaling**: The same is true for scaling the digits so that they all have the same size. This way, the location of strikes, curves, and loops are important. Otherwise, the neural network might easily confuse eights and zeros because they are all made of one or two closed loops. Fortunately, images in MNIST are already scaled.
- **Representing categorical features**: It is important that target labels be one-hot encoded so that we can have ten neurons in the output layer corresponding to the ten different classes 0-9. This step we still have to perform ourselves.

The easiest way to transform `y_train` and `y_test` is by the one-hot encoder from scikit-learn:

```
In [6]: from sklearn.preprocessing import OneHotEncoder
   ...: enc = OneHotEncoder(sparse=False, dtype=np.float32)
   ...: y_train_pre = enc.fit_transform(y_train.reshape(-1, 1))
```

This will transform the labels of the training set from a `<n_samples x 1>` vector with integers 0-9 into a `<n_samples x 10>` matrix with floating point numbers 0.0 or 1.0. Analogously, we can transform `y_test` using the same procedure:

```
In [7]: y_test_pre = enc.fit_transform(y_test.reshape(-1, 1))
```

In addition, we need to preprocess `X_train` and `X_test` for the purpose of working with OpenCV. Currently, `X_train` and `X_test` are 3-D matrices `<n_samples x 28 x 28>` with integer values between 0 and 255. Preferably, we want a 2-D matrix `<n_samples x n_features>` with floating point numbers, where `n_features` is 784:

```
In [8]: X_train_pre = X_train_pre.reshape((X_train.shape[0], -1))
   ...: X_train_pre = X_train.astype(np.float32) / 255.0
   ...: X_test_pre = X_test_pre.reshape((X_test.shape[0], -1))
   ...: X_test_pre = X_test.astype(np.float32) / 255.0
```

Then we are ready to train the network.

Training an MLP using OpenCV

We can set up and train an MLP in OpenCV with the following recipe:

1. Instantiate a new MLP object:

   ```
   In [9]: import cv2
      ...: mlp = cv2.ml.ANN_MLP_create()
   ```

2. Specify the size of every layer in the network. We are free to add as many layers as we want, but we need to make sure that the first layer has the same number of neurons as input features (784 in our case), and that the last layer has the same number of neurons as class labels (10 in our case):

   ```
   In [10]: mlp.setLayerSizes(np.array([784, 512, 512, 10]))
   ```

3. Specify an activation function. Here we use the sigmoidal activation function from before:

   ```
   In [11]: mlp.setActivationFunction(cv2.ml.ANN_MLP_SIGMOID_SYM,
      ...:                            2.5, 1.0)
   ```

4. Specify the training method. Here we use the backpropagation algorithm described above. We also need to make sure that we choose a small enough learning rate. Since we have on the order of 10^5 training samples, it is a good idea to set the learning rate to at most 10^{-5}:

   ```
   In [12]: mlp.setTrainMethod(cv2.ml.ANN_MLP_BACKPROP)
      ...:  mlp.setBackpropWeightScale(0.00001)
   ```

5. Specify the termination criteria. Here we use the same criteria as above: to run training for ten iterations (`term_max_iter`) or until the error does no longer increase significantly (`term_eps`):

   ```
   In [13]: term_mode = (cv2.TERM_CRITERIA_MAX_ITER +
      ...:               cv2.TERM_CRITERIA_EPS)
      ...:  term_max_iter = 10
      ...:  term_eps = 0.01
      ...:  mlp.setTermCriteria((term_mode, term_max_iter,
      ...:                       term_eps))
   ```

5. Train the network on the training set (`X_train_pre`):

```
In [14]: mlp.train(X_train_pre, cv2.ml.ROW_SAMPLE, y_train_pre)
Out[14]: True
```

Before you call `mlp.train`, here is a word of caution: this might take several hours to run, depending on your computer setup! For comparison, it took just under an hour on my own laptop. We are now dealing with a real-world dataset of 60,000 samples: if we run 100 training epochs, we have to compute 6 million gradients! So beware.

When the training completes, we can calculate the accuracy score on the training set to see how far we got:

```
In [15]: _, y_hat_train = mlp.predict(X_train_pre)
In [16]: from sklearn.metrics import accuracy_score
   ...:  accuracy_score(y_hat_train.round(), y_train_pre)
Out[16]: 0.92976666666666663
```

But, of course, what really counts is the accuracy score we get on the held-out test data:

```
In [17]: _, y_hat_test = mlp.predict(X_test_pre)
   ...:  accuracy_score(y_hat_test.round(), y_test_pre)
Out[17]: 0.91690000000000005
```

91.7% accuracy is not bad at all if you ask me! The first thing you should try is to change the layer sizes in `In [10]` above, and see how the test score changes. As you add more neurons to the network, you should see the training score increase—and with it, hopefully, the test score. However, having N neurons in a single layer is not the same as having them spread out over several layers! Can you confirm this observation?

Training a deep neural net using Keras

Although we achieved a formidable score with the MLP above, our result does not hold up to state-of-the-art results. Currently the best result has close to 99.8% accuracy—better than human performance! This is why nowadays, the task of classifying handwritten digits is largely regarded as solved.

To get closer to the state-of-the-art results, we need to use state-of-the-art techniques. Thus, we return to Keras.

Preprocessing the MNIST dataset

To make sure we get the same result every time we run the experiment, we will pick a random seed for NumPy's random number generator. This way, shuffling the training samples from the MNIST dataset will always result in the same order:

```
In [1]: import numpy as np
   ...      np.random.seed(1337)
```

Keras provides a loading function similar to `train_test_split` from scikit-learn's `model_selection` module. Its syntax might look strangely familiar to you:

```
In [2]: from keras.datasets import mnist
   ...      (X_train, y_train), (X_test, y_test) = mnist.load_data()
```

In contrast to other datasets we have encountered so far, MNIST comes with a pre-defined train-test split. This allows the dataset to be used as a **benchmark**, as the test score reported by different algorithms will always apply to the same test samples.

The neural nets in Keras act on the feature matrix slightly differently than the standard OpenCV and scikit-learn estimators. Whereas the rows of a feature matrix in Keras still correspond to the number of samples (`X_train.shape[0]` in the code below), we can preserve the two-dimensional nature of the input images by adding more dimensions to the feature matrix:

```
In [3]: img_rows, img_cols = 28, 28
   ...      X_train = X_train.reshape(X_train.shape[0], img_rows, img_cols, 1)
   ...      X_test = X_test.reshape(X_test.shape[0], img_rows, img_cols, 1)
   ...      input_shape = (img_rows, img_cols, 1)
```

Here we have reshaped the feature matrix into a four-dimensional matrix with dimensions `n_features x 28 x 28 x 1`. We also need to make sure we operate on 32-bit floating point numbers between [0, 1], rather than unsigned integers in [0, 255]:

```
   ...      X_train = X_train.astype('float32') / 255.0
   ...      X_test = X_test.astype('float32') / 255.0
```

Then, we can one-hot encode the training labels like we did before. This will make sure each category of target labels can be assigned to a neuron in the output layer. We could do this with scikit-learn's `preprocessing`, but in this case it is easier to use Keras' own utility function:

```
In [4]: from keras.utils import np_utils
   ...      n_classes = 10
   ...      Y_train = np_utils.to_categorical(y_train, n_classes)
```

```
...         Y_test = np_utils.to_categorical(y_test, n_classes)
```

Creating a convolutional neural network

Once we have preprocessed the data, it is time to define the actual model. Here, we will once again rely on the `Sequential` model to define a feedforward neural network:

```
In [5]: from keras.model import Sequential
...     model = Sequential()
```

However, this time, we will be smarter about the individual layers. We will design our neural network around a **convolutional layer**, where the kernel is a *3 x 3* pixel two-dimensional convolution:

```
In [6]: from keras.layers import Convolution2D
...     n_filters = 32
...     kernel_size = (3, 3)
...     model.add(Convolution2D(n_filters, kernel_size[0], kernel_size[1],
...                             border_mode='valid',
...                             input_shape=input_shape))
```

A two-dimensional convolutional layer operates akin to image filtering in OpenCV, where each image in the input data is convolved with a small two-dimensional kernel. In Keras, we can specify the **kernel size** and the **stride**. For more information on convolutional layers, I can recommend the following book: John Hearty (2016). *Advanced Machine Learning with Python*. Packt Publishing, ISBN 978-1-78439-863-7.

After that, we will use a linear rectified unit as an activation function:

```
In [7]: from keras.layers import Activation
...     model.add(Activation('relu'))
```

In a deep convolutional neural net, we can have as many layers as we want. A popular version of this structure applied to MNIST involves performing the convolution and rectification twice:

```
In [8]: model.add(Convolution2D(n_filters, kernel_size[0], kernel_size[1]))
...     model.add(Activation('relu'))
```

Finally, we will pool the activations and add a Dropout layer:

```
In [9]: from keras.layers import MaxPooling2D, Dropout
   ...  pool_size = (2, 2)
   ...  model.add(MaxPooling2D(pool_size=pool_size))
   ...  model.add(Dropout(0.25))
```

Then we will flatten the model and finally pass it through a `softmax` function to arrive at the output layer:

```
In [10]: from keras.layers import Flatten, Dense
   ...   model.add(Flatten())
   ...   model.add(Dense(128))
   ...   model.add(Activation('relu'))
   ...   model.add(Dropout(0.5))
   ...   model.add(Dense(n_classes))
   ...   model.add(Activation('softmax'))
```

Here, we will use the cross-entropy loss and the **Adadelta** algorithm:

```
In [11]: model.compile(loss='categorical_crossentropy',
   ...                 optimizer='adadelta',
   ...                 metrics=['accuracy'])
```

Fitting the model

We fit the model like we do with all other classifiers (caution, this might take a while):

```
In [12]: model.fit(X_train, Y_train, batch_size=128, nb_epoch=12,
   ...             verbose=1, validation_data=(X_test, Y_test))
```

After training completes, we can evaluate the classifier:

```
In [13]: model.evaluate(X_test, Y_test, verbose=0)
Out[13]: 0.9925
```

And we achieve 99.25% accuracy! Worlds apart from the MLP classifier we implemented before. And this is just one way to do things. As you can see, neural networks provide a plethora of tuning parameters, and it is not at all clear which ones will lead to the best performance.

Summary

In this chapter, we added a whole bunch of skills to our list as a machine learning practitioner. Not only did we cover the basics of artificial neural networks, including the perceptron and multilayer perceptrons, we also got our hands on some advanced deep learning software. We learned how to build a simple perceptron from scratch and how to build state-of-the-art networks using Keras. Furthermore, we learnt about all the details of neural nets: activation functions, loss functions, layer types, and training methods. All in all, this was probably the densest chapter yet.

Now that you know about most of the essential supervised learners, it is time to talk about how to combine different algorithms into a more powerful one. Thus, in the next chapter, we will talk about how to build ensemble classifiers.

10
Combining Different Algorithms into an Ensemble

So far, we have looked at a number of interesting machine learning algorithms, from classic methods such as linear regression to more advanced techniques such as deep neural networks. At various points, we pointed out that every algorithm has its own strengths and weaknesses—and we took note of how to spot and overcome these weaknesses.

However, wouldn't it be great if we could simply stack together a bunch of average classifiers to form a much stronger **ensemble** of classifiers?

In this chapter, we will do just that. Ensemble methods are techniques that bind multiple different models together in order to solve a shared problem. Their use has become a common practice in competitive machine learning—making use of an ensemble typically improves an individual classifier's performance by a small percentage.

Among these techniques are the so-called **bagging methods**, where the vote of multiple classifiers is averaged to make a final decision, and **boosting methods**, where one classifier is trying to rectify the errors made by another. One of these methods is known as **random forest**, which is a combination of multiple decision trees. In addition, you might already be familiar with **Adaptive Boosting** (also known as **AdaBoost**), a powerful boosting technique and popular feature of OpenCV.

As we progress through the chapter, we want to find answers to the following questions:

- How can we combine multiple models to form an ensemble classifier?
- What are random forests and what do they have to do with decision trees?
- What's the different between bagging and boosting?

Let's jump right in!

Understanding ensemble methods

The goal of ensemble methods is to combine the predictions of several individual estimators built with a given learning algorithm in order to solve a shared problem. Typically, an ensemble consists of two major components:

- a set of models
- a set of **decision rules** that govern how the results of these models are combined into a single output

The idea behind ensemble methods has much to do with the **wisdom of the crowd** concept. Rather than the opinion of a single expert, we consider the collective opinion of a group of individuals. In the context of machine learning, these individuals would be **classifiers** or **regressors**. The idea is that if we just ask a large enough number of classifiers, one of them ought to get it right.

A consequence of this procedure is that we get a multitude of opinions about any given problem. So how do we know which classifier is right?

This is why we need a decision rule. Perhaps we consider everybody's opinion of equal importance, or perhaps we would want to weight somebody's opinion based on their expert status. Depending on the nature of our decision rule, ensemble methods can be categorized as follows:

- **Averaging methods**: They develop models in parallel and then use averaging or voting techniques to come up with a combined estimator. This is as close to democracy as ensemble methods can get.
- **Boosting methods**: They involve building models in sequence, where each added model aims to improve the score of the combined estimator. This is akin to debugging the code of your intern or reading the report of your undergraduate student: they are all bound to make errors, and the job of every subsequent expert laying eyes on the topic is to figure out the special cases where the preceding expert got it wrong.
- **Stacking methods**: Also known as **blending methods**, they use the weighted output of multiple classifiers as inputs to the next layer in the model. This is akin to having expert groups who pass on their decision to the next expert group.

The same three categories of ensemble methods are illustrated in the following figure:

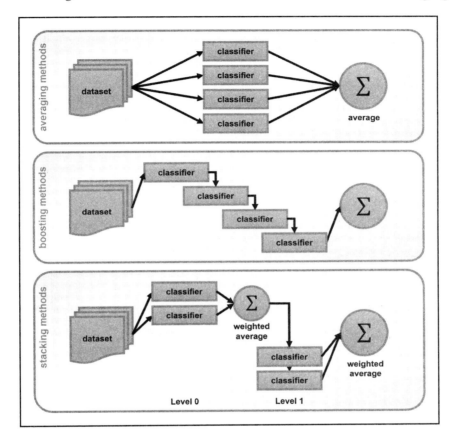

Ensemble methods

Ensemble classifiers afford us the opportunity to choose the best elements of each of the proposed solutions in order to come up with a single, final result. As a result, we typically bet much more accurate and robust results.

 When building an ensemble classifier, we should be mindful that our goal is to tune the performance of the ensemble rather than of the models that comprise it. Individual model performance is thus less important than the overall performance of the ensemble.

Let's briefly discuss the three different types of ensemble methods and talk about how they can be used to classify data.

Understanding averaging ensembles

Averaging methods have a long history in machine learning and are commonly applied in fields such as molecular dynamics and audio signal processing. Such ensembles are typically seen as almost exact replicates of a given system.

An **averaging ensemble** is essentially a collection of models that train on the same dataset. Their results are then aggregated in a number of ways.

One common method involves creating multiple model configurations that take different parameter subsets as input. Techniques that take this approach are referred to collectively as bagging methods.

Bagging methods come in many different flavors. However, they typically only differ by the way they draw random subsets of the training set:

- **Pasting** methods draw random subsets of the samples without replacement
- **Bagging** methods draw random subsets of the samples with replacement
- **Random subspace** methods draw random subsets of the features but train on all data samples
- **Random patches** methods draw random subsets of both samples and features

Averaging ensembles can be used to reduce the variability of a model's performance.

In scikit-learn, bagging methods can be realized using the **meta-estimators** `BaggingClassifier` and `BaggingRegressor`. These are meta-estimators because they allow us to build an ensemble from any other base estimator.

Implementing a bagging classifier

We can, for instance, build an ensemble from a collection of 10 k-NN classifiers as follows:

```
In [1]: from sklearn.ensemble import BaggingClassifier
   ...: from sklearn.neighbors import KNeighborsClassifier
   ...: bag_knn = BaggingClassifier(KNeighborsClassifier(),
   ...:                             n_estimators=10)
```

The `BaggingClassifier` class provides a number of options to customize the ensemble:

- `n_estimators`: As shown in the preceding code, this specifies the number of base estimators in the ensemble.
- `max_samples`: This denotes the number (or fraction) of samples to draw from the dataset to train each base estimator. We can set `bootstrap=True` to sample with replacement (effectively implementing bagging), or we can set `bootstrap=False` to implement pasting.
- `max_features`: This denotes the number (or fraction) of features to draw from the feature matrix to train each base estimator. We can set `max_samples=1.0` and `max_features<1.0` to implement the random subspace method. Alternatively, we can set both `max_samples<1.0` and `max_features<1.0` to implement the random patches method.

This gives us the ultimate freedom to implement any kind of averaging ensemble. An ensemble can then be fit to data like any other estimator using the ensemble's `fit` method.

For example, if we wanted to implement bagging with 10 k-NN classifiers with k=5, where every k-NN classifier is trained on 50% of the samples in the dataset, we would modify the preceding command as follows:

```
In [2]: bag_knn = BaggingClassifier(KNeighborsClassifier(n_neighbors=5),
   ...                              n_estimators=10, max_samples=0.5,
   ...                              bootstrap=True, random_state=3)
```

In order to observe a performance boost, we have to apply the ensemble to some dataset, such as the breast cancer dataset from Chapter 5, *Using Decision Trees to Make a Medical Diagnosis*:

```
In [3]: from sklearn.datasets import load_breast_cancer
   ...  dataset = load_breast_cancer()
   ...  X = dataset.data
   ...  y = dataset.target
```

As usual, we should follow the best practice of splitting the data into training and test sets:

```
In [4]: from sklearn.model_selection import train_test_split
   ...  X_train, X_test, y_train, y_test = train_test_split(
   ...      X, y, random_state=3
   ...  )
```

Then we can train the ensemble using the `fit` method and evaluate its generalization performance using the `score` method:

```
In [5]: bag_knn.fit(X_train, y_train)
   ...: bag_knn.score(X_test, y_test)
Out[5]: 0.93706293706293708
```

The performance boost will become evident once we also train a single *k*-NN classifier on the data:

```
In [6]: knn = KNeighborsClassifier(n_neighbors=5)
   ...: knn.fit(X_train, y_train)
   ...: knn.score(X_test, y_test)
Out[6]: 0.91608391608391604
```

Without changing the underlying algorithm, we were able to improve our test score from 91.6% to 93.7% by simply letting 10 *k*-NN classifiers do the job instead of a single one.

You're welcome to experiment with other bagging ensembles. For example, how would you adjust the preceding code snippets to implement the random patches method?

 You can find the answer in the Jupyter Notebook `notebooks/10.00-Combining-Different-Algorithms-Into-an-Ensemble.ipynb` on GitHub.

Implementing a bagging regressor

Similarly, we can use the `BaggingRegressor` class to form an ensemble of regressors.

For example, we could build an ensemble of decision trees to predict the housing prices from the Boston dataset of Chapter 3, *First Steps in Supervised Learning*. The syntax is almost identical to setting up a bagging classifier:

```
In [7]: from sklearn.ensemble import BaggingRegressor
   ...: from sklearn.tree import DecisionTreeRegressor
   ...: bag_tree = BaggingRegressor(DecisionTreeRegressor(),
   ...:                             max_features=0.5, n_estimators=10,
   ...:                             random_state=3)
```

Of course, we need to load and split the dataset like we did for the breast cancer dataset:

```
In [8]: from sklearn.datasets import load_boston
   ...: dataset = load_boston()
   ...: X = dataset.data
   ...: y = dataset.target
```

```
In [9]: from sklearn.model_selection import train_test_split
   ...: X_train, X_test, y_train, y_test = train_test_split(
   ...:     X, y, random_state=3
   ...: )
```

Then we can fit the bagging regressor on `X_train` and score it on `X_test`:

```
In [10]: bag_tree.fit(X_train, y_train)
   ...:  bag_tree.score(X_test, y_test)
Out[10]: 0.82704756225081688
```

As in the preceding example, we find a performance boost of roughly 5%, from 77.3% accuracy for a single decision tree to 82.7% accuracy.

Of course, we wouldn't just stop here. Nobody said the ensemble needs to consist of 10 individual estimators, so we are free to explore different-sized ensembles. On top of that, the `max_samples` and `max_features` parameters allow for a great deal of customization.

A more sophisticated version of bagged decision trees is called random forests, which we will talk about later in this chapter.

Understanding boosting ensembles

Another approach to building ensembles is through boosting. Boosting models use multiple individual learners in sequence to iteratively **boost** the performance of the ensemble.

Typically, the learners used in boosting are relatively simple. A good example is a decision tree with only a single node—a **decision stump**. Another example could be a simple linear regression model. The idea is not to have the strongest individual learners, quite the opposite—we want the individuals to be **weak learners**, so that we get a superior performance only when we consider a large number of individuals.

At each iteration of the procedure, the training set is adjusted so that the next classifier is applied to the data points that the preceding classifier got wrong. Over multiple iterations, the ensemble is extended with a new tree (whichever tree optimized the ensemble performance score) at each iteration.

In scikit-learn, boosting methods can be realized using the `GradientBoostingClassifier` and `GradientBoostingRegressor` objects.

Implementing a boosting classifier

For example, we can build a boosting classifier from a collection of 10 decision trees as follows:

```
In [11]: from sklearn.ensemble import GradientBoostingClassifier
    ...      boost_class = GradientBoostingClassifier(n_estimators=10,
    ...                                               random_state=3)
```

These classifiers support both binary and multiclass classification.

Similar to the `BaggingClassifier` class, the `GradientBoostingClassifier` class provides a number of options to customize the ensemble:

- `n_estimators`: This denotes the number of base estimators in the ensemble. A large number of estimators typically results in better performance.
- `loss`: This denotes the loss function (or cost function) to be optimized. Setting `loss='deviance'` implements logistic regression for classification with probabilistic outputs. Setting `loss='exponential'` actually results in AdaBoost, which we will talk about in a little bit.
- `learning_rate`: This denotes the fraction by which to shrink the contribution of each tree. There is a trade-off between `learning_rate` and `n_estimators`.
- `max_depth`: This denotes the maximum depth of the individual trees in the ensemble.
- `criterion`: This denotes the function to measure the quality of a node split.
- `min_samples_split`: This denotes the number of samples required to split an internal node.
- `max_leaf_nodes`: This denotes the maximum number of leaf nodes allowed in each individual tree and so on.

We can apply the boosted classifier to the preceding breast cancer dataset to get an idea of how this ensemble compares to a bagged classifier. But first, we need to reload the dataset:

```
In [12]: dataset = load_breast_cancer()
    ...      X = dataset.data
    ...      y = dataset.target
In [13]: X_train, X_test, y_train, y_test = train_test_split(
    ...          X, y, random_state=3
    ...      )
```

Then we find that the boosted classifier achieves 94.4% accuracy on the test set—a little under 1% better than the preceding bagged classifier:

```
In [14]: boost_class.fit(X_train, y_train)
    ...  boost_class.score(X_test, y_test)
Out[14]: 0.94405594405594406
```

We would expect an even better score if we increased the number of base estimators from 10 to 100. In addition, we might want to play around with the learning rate and the depths of the trees.

Implementing a boosting regressor

Implementing a boosted regressor follows the same syntax as the boosted classifier:

```
In [15]: from sklearn.ensemble import GradientBoostingRegressor
    ...  boost_reg = GradientBoostingRegressor(n_estimators=10,
    ...                                       random_state=3)
```

We have seen earlier that a single decision tree can achieve 79.3% accuracy on the Boston dataset. A bagged decision tree classifier made of 10 individual regression trees achieved 82.7% accuracy. But how does a boosted regressor compare?

Let's reload the Boston dataset and split it into training and test sets. We want to make sure we use the same value for `random_state` so that we end up training and testing on the same subsets of the data:

```
In [16]: dataset = load_boston()
    ...  X = dataset.data
    ...  y = dataset.target
In [17]: X_train, X_test, y_train, y_test = train_test_split(
    ...      X, y, random_state=3
    ...  )
```

As it turns out, the boosted decision tree ensemble actually performs worse than the previous code:

```
In [18]: boost_reg.fit(X_train, y_train)
    ...  boost_reg.score(X_test, y_test)
Out[18]: 0.71991199075668488
```

This result might be confusing at first. After all, we used 10 times more classifiers than we did for the single decision tree. Why would our numbers get worse?

You can see, this is a good example of an expert classifier being smarter than a group of weak learners. One possible solution is to make the ensemble larger. In fact, it is customary to use on the order of 100 weak learners in a boosted ensemble:

```
In [19]: boost_reg = GradientBoostingRegressor(n_estimators=100)
```

Then, when we retrain the ensemble on the Boston dataset, we get a test score of 89.8%:

```
In [20]: boost_reg.fit(X_train, y_train)
    ...  boost_reg.score(X_test, y_test)
Out[20]: 0.89984081091774459
```

What happens when you increase the number to `n_estimators=500`? There's a lot more we could do by playing with the optional parameters.

As you can see, boosting is a powerful procedure that allows you to get massive performance improvements by combining a large number of relatively simple learners.

A specific implementation of boosted decision trees is the AdaBoost algorithm, which we will talk about later in this chapter.

Understanding stacking ensembles

All the ensemble methods we have seen so far share a common design philosophy: to fit multiple individual classifiers to the data and incorporate their predictions with the help of some simple decision rule (such as averaging or boosting) into a final prediction.

Stacking ensembles, on the other hand, build ensembles with hierarchies. Here, individual learners are organized into multiple layers where the output of one layer of learners is used as training data for a model at the next layer. This way, it is possible to successfully blend hundreds of different models.

Unfortunately, we won't have the time to discuss stacking ensembles in detail.

However, these models can be very powerful, as seen, for example, in the Netflix Prize competition. The most successful models in the competition used to include stacking ensembles that not just combined different layers but extracted from each layer the parameters that were most effective and used them as meta-features for training the next layer.

You can learn more about the Netflix Prize at http://www.netflixprize.com. One of the stacking models that was used to achieve the second prize in the competition is also described in the following scientific article: Joseph Sill, Gabor Takacs, Lester Mackey, David Lin (2009). Feature-Weighted Linear Stacking, https://arxiv.org/abs/0911.0460.

Let's now move on to discuss some popular bagging and boosting techniques, such as random forests and Adaptive Boosting.

Combining decision trees into a random forest

A popular variation of bagged decision trees are the so-called random forests. These are essentially a collection of decision trees, where each tree is slightly different from the others. In contrast to bagged decision trees, each tree in a random forest is trained on a slightly different subset of data features.

Although a single tree of unlimited depth might do a relatively good job of predicting the data, it is also prone to overfitting. The idea behind random forests is to build a large number of trees, each of them trained on a random subset of data samples and features. Because of the randomness of the procedure, each tree in the forest will overfit the data in a slightly different way. The effect of overfitting can then be reduced by averaging the predictions of the individual trees.

Understanding the shortcomings of decision trees

The effect of overfitting the dataset, which a decision tree often falls victim of is best demonstrated through a simple example.

For this, we will return to the `make_moons` function from scikit-learn's `datasets` module, which we previously used in Chapter 8, *Discovering Hidden Structures with Unsupervised Learning* to organize data into two interleaving half circles. Here, we choose to generate 100 data samples belonging to two half circles, in combination with some Gaussian noise with standard deviation 0.25:

```
In [1]: from sklearn.datasets import make_moons
   ...      X, y = make_moons(n_samples=100, noise=0.25,
   ...                        random_state=100)
```

We can visualize this data using Matplotlib and the `scatter` function:

```
In [2]: import matplotlib.pyplot as plt
   ...     %matplotlib inline
   ...     plt.style.use('ggplot')
In [3]: plt.scatter(X[:, 0], X[:, 1], s=100, c=y)
   ...     plt.xlabel('feature 1')
   ...     plt.ylabel('feature 2');
```

The resulting plot has the data colored according to the class they belong to:

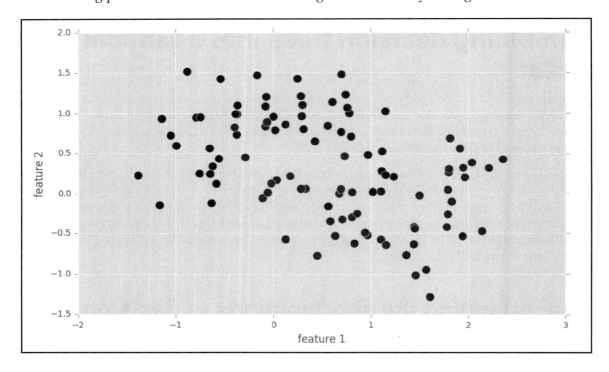

Example of a `make_moons` dataset with added noise

Because of all the noise we added, the two half moons might not be apparent at first glance. That's a perfect scenario for our current intentions, which is to show that decision trees are tempted to overlook the general arrangement of data points (that is, the fact that they are organized in half circles) and instead focus on the noise in the data.

To illustrate this point, we first need to split the data into training and test sets. We choose a comfortable 75-25 split (by not specifying `train_size`), as we have done a number of times before:

```
In [4]: from sklearn.model_selection import train_test_split
   ...:     X_train, X_test, y_train, y_test = train_test_split(
   ...:         X, y, random_state=100
   ...:     )
```

Now let's have some fun. What we want to do is to study how the decision boundary of a decision tree changes as we make it deeper and deeper.

For this, we will bring back the `plot_decision_boundary` function from Chapter 6, *Detecting Pedestrians with Support Vector Machines* among others:

```
In [5]: import numpy as np
   ...: def plot_decision_boundary(classifier, X_test, y_test):
```

The function itself involves a number of processing steps:

1. Create a dense 2-D grid of data points between (x_min, y_min), and (x_max, y_max):

    ```
    ...             h = 0.02 # step size in mesh
    ...             x_min = X_test[:, 0].min() - 1,
    ...             x_max = X_test[:, 0].max() + 1
    ...             y_min = X_test[:, 1].min() - 1
    ...             y_max = X_test[:, 1].max() + 1
    ...             xx, yy = np.meshgrid(np.arange(x_min, x_max, h),
    ...                                  np.arange(y_min, y_max, h))
    ...
    ```

2. Classify every point on the grid:

    ```
    ...             X_hypo = np.c_[xx.ravel().astype(np.float32),
    ...                            yy.ravel().astype(np.float32)]
    ...             zz = classifier.predict(X_hypo)
    ...             zz = zz.reshape(xx.shape)
    ```

3. If the `predict` method returns a tuple, we are dealing with an OpenCV classifier, else we are dealing with scikit-learn:

    ```
    ...             if isinstance(ret, tuple):
    ...                 zz = ret[1]
    ...             else:
    ...                 zz = ret
    ...             zz = zz.reshape(xx.shape)
    ```

4. Use the predicted target labels (`zz`) to color the decision landscape:

```
...         plt.contourf(xx, yy, zz, cmap=plt.cm.coolwarm,
                         alpha=0.8)
```

5. Draw all the data points from the test set on top of the colored landscape:

```
...         plt.scatter(X_test[:, 0], X_test[:, 1], c=y_test,
...                     s=200)
```

Then we can code up a for loop, where at each iteration, we fit a tree of a different depth:

```
In [6]: from sklearn.tree import DecisionTreeClassifier
   ...     for depth in range(1, 9):
```

In order to plot the decision landscape for every created decision tree, we need to create a plot with a total of eight subplots. The iterator `depth` can then serve as an index for the current subplot:

```
...         plt.subplot(2, 4, depth)
```

After fitting the tree to the data, we pass the predicted test labels to the `plot_decision_boundary` function:

```
...         tree = DecisionTreeClassifier(max_depth=depth)
...         tree.fit(X_train, y_train)
...         y_test = tree.predict(X_test)
...         plot_decision_boundary(tree, X_test, y_test)
```

For the sake of clarity, we also turn off the axes and decorate each plot with a title:

```
...         plt.axis('off')
...         plt.title('depth = %d' % depth)
```

This produces the following plot:

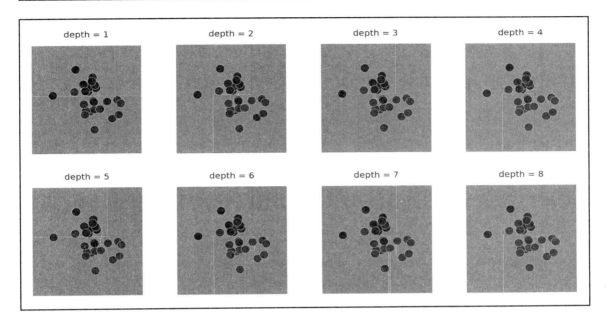

Classification results for decision trees with different depths

Recall the procedure of a decision tree: at every node in the tree, the algorithm picks a feature to split, either along the *x*-axis or along the *y*-axis. For example, a tree of `depth=1` has only one node and, thus, makes only one decision. This particular tree (upper row, leftmost panel) chose to split the second feature right down the middle, leading to a horizontal decision boundary that separates the decision landscape into an upper and lower half.

Similarly, a decision tree of `depth=2` will go a step further. After making the first decision of splitting the second feature in layer #1, the algorithm ends up with two additional nodes in layer #2. This particular tree (upper row, second panel from the left) chose to split the first feature in layer #2, leading to two additional vertical lines.

As we continue to build deeper and deeper trees, we notice something strange: the deeper the tree, the more likely it is to get strangely shaped decision regions, such as the tall and skinny patches in the rightmost panel of the lower row. It's clear that these patches are more a result of the noise in the data rather than some characteristic of the underlying data distribution. This is an indication that most of the trees are overfitting the data. After all, we know for a fact that the data is organized into two half circles! As such, the trees with `depth=3` or `depth=5` are probably closest to the real data distribution.

There are at least two different ways to make a decision tree less powerful:

- Train the tree only on a subset of the data
- Train the tree only on a subset of the features

Random forests do just that. In addition, they repeat the experiment many times by building an ensemble of trees, each of which is trained on a randomly chosen subset of data samples and/or features.

Random forests are a powerful ensemble classifier that come with several advantages:

- Both training and evaluation are fast. Not only are the underlying decision trees relatively simple data structures, but every tree in the forest is independent, making it easy to train them in parallel.
- Multiple trees allow for probabilistic classification. Later in this chapter, we will talk about voting procedures, which allow an ensemble to predict the probability with which a data point belongs to a particular class.

That being said, one of the disadvantages of random forests is that they might be hard to interpret, and the same goes for individual trees that are relatively deep. Although cumbersome, it is still possible to investigate the decision path within a particular tree. You might want to check out Chapter 5, *Using Decision Trees to Make a Medical Diagnosis* for a refresher.

Implementing our first random forest

In OpenCV, random forests can be built using the `RTrees_create` function from the `ml` module:

```
In [7]: import cv2
   ...:     rtree = cv2.ml.RTrees_create()
```

The tree object provides a number of options, the most important of which are the following:

- `setMaxDepth`: This sets the maximum possible depth of each tree in the ensemble. The actual obtained depth may be smaller if other termination criteria are met first.
- `setMinSampleCount`: This sets the minimum number of samples that a node can contain for it to get split.

- `setMaxCategories`: This sets the maximum number of categories allowed. Setting the number of categories to a smaller value than the actual number of classes in the data performs subset estimation.
- `setTermCriteria`: This sets the termination criteria of the algorithm. This is also where you set the number of trees in the forest.

> Although we might have hoped for a `setNumTrees` method to set the number of trees in the forest (kind of the most important parameter of them all, no?), we instead need to rely on the `setTermCriteria` method. Confusingly, the number of trees is conflated with `cv2.TERM_CRITERA_MAX_ITER`, which is usually reserved for the number of iterations that an algorithm is run for, not for the number of estimators in an ensemble.

We can specify the number of trees in the forest by passing an integer `n_trees` to the `setTermCriteria` method. Here, we also want to tell the algorithm to quit once the score does not increase by at least `eps` from one iteration to the next:

```
In [8]: n_trees = 10
   ...: eps = 0.01
   ...: criteria = (cv2.TERM_CRITERIA_MAX_ITER + cv2.TERM_CRITERIA_EPS,
   ...:             n_trees, eps)
   ...: rtree.setTermCriteria(criteria)
```

Then we are ready to train the classifier on the data from the preceding code:

```
In [9]: rtree.train(X_train.astype(np.float32), cv2.ml.ROW_SAMPLE,
                    y_train);
```

The test labels can be predicted with the `predict` method:

```
In [10]: _, y_hat = rtree.predict(X_test.astype(np.float32))
```

Using scikit-learn's `accuracy_score`, we can evaluate the model on the test set:

```
In [11]: from sklearn.metrics import accuracy_score
    ...: accuracy_score(y_test, y_hat)
Out[11]: 0.83999999999999997
```

After training, we can pass the predicted labels to the `plot_decision_boundary` function:

```
In [12]: plot_decision_boundary(rtree, X_test, y_test)
```

This will produce the following plot:

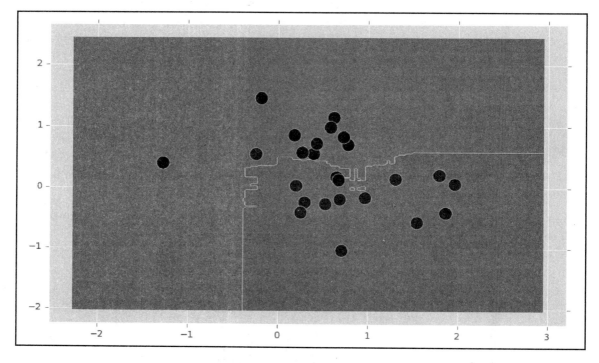

Decision landscape of a random forest classifier

Implementing a random forest with scikit-learn

Alternatively, we can implement random forests using scikit-learn:

```
In [13]: from sklearn.ensemble import RandomForestClassifier
    ...:     forest = RandomForestClassifier(n_estimators=10, random_state=200)
```

Here, we have a number of options to customize the ensemble:

- n_estimators: This specifies the number of trees in the forest.
- criterion: This specifies the node splitting criterion. Setting criterion='gini' implements the Gini impurity, whereas setting criterion='entropy' implements information gain.

- `max_features`: This specifies the number (or fraction) of features to consider at each node split.
- `max_depth`: This specifies the maximum depth of each tree.
- `min_samples`: This specifies the minimum number of samples required to split a node.

We can then fit the random forest to the data and score it like any other estimator:

```
In [14]: forest.fit(X_train, y_train)
    ...  forest.score(X_test, y_test)
Out[14]: 0.83999999999999997
```

This gives roughly the same result as in OpenCV. We can use our helper function to plot the decision boundary:

```
In [15]: plot_decision_boundary(forest, X_test, y_test)
```

The output looks like this:

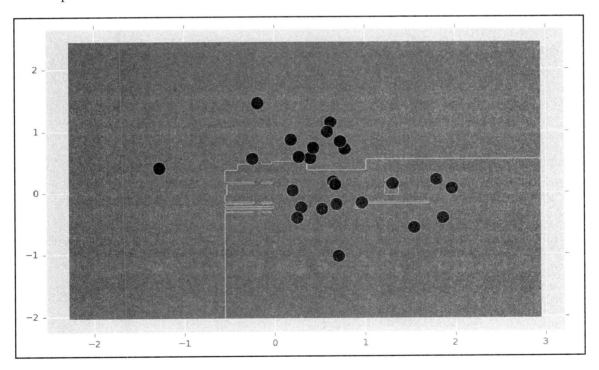

Decision boundary of a random forest

Implementing extremely randomized trees

Random forests are already pretty arbitrary. But what if we wanted to take the randomness to its extreme?

In **extremely randomized trees** (see `ExtraTreesClassifier` and `ExtraTreesRegressor` classes), the randomness is taken even further than in random forests. Remember how decision trees usually choose a threshold for every feature so that the purity of the node split is maximized. Extremely randomized trees, on the other hand, choose these thresholds at random. The best one of these randomly-generated thresholds is then used as the splitting rule.

We can build an extremely randomized tree as follows:

```
In [16]: from sklearn.ensemble import ExtraTreesClassifier
...      extra_tree = ExtraTreesClassifier(n_estimators=10, random_state=100)
```

To illustrate the difference between a single decision tree, a random forest, and extremely randomized trees, let's consider a simple dataset, such as the Iris dataset:

```
In [17]: from sklearn.datasets import load_iris
...      iris = load_iris()
...      X = iris.data[:, [0, 2]]
...      y = iris.target
In [18]: X_train, X_test, y_train, y_test = train_test_split(
...          X, y, random_state=100
...      )
```

We can then fit and score the tree object the same way we did before:

```
In [19]: extra_tree.fit(X_train, y_train)
...      extra_tree.score(X_test, y_test)
Out[19]: 0.92105263157894735
```

For comparison, using a random forest would have resulted in the same performance:

```
In [20]: forest = RandomForestClassifier(n_estimators=10,
                                          random_state=100)
...      forest.fit(X_train, y_train)
...      forest.score(X_test, y_test)
Out[20]: 0.92105263157894735
```

In fact, the same is true for a single tree:

```
In [21]: tree = DecisionTreeClassifier()
...      tree.fit(X_train, y_train)
```

```
...            tree.score(X_test, y_test)
Out[21]: 0.92105263157894735
```

So what's the difference between them? To answer this question, we have to look at the decision boundaries. Fortunately, we have already imported our `plot_decision_boundary` helper function in the preceding section, so all we need to do is pass the different classifier objects to it.

We will build a list of classifiers, where each entry in the list is a tuple that contains an index, a name for the classifier, and the classifier object:

```
In [22]: classifiers = [
...          (1, 'decision tree', tree),
...          (2, 'random forest', forest),
...          (3, 'extremely randomized trees', extra_tree)
...      ]
```

Then it's easy to pass the list of classifiers to our helper function such that the decision landscape of every classifier is drawn in its own subplot:

```
In [23]: for sp, name, model in classifiers:
...          plt.subplot(1, 3, sp)
...          plot_decision_boundary(model, X_test, y_test)
...          plt.title(name)
...          plt.axis('off')
```

The result looks like this:

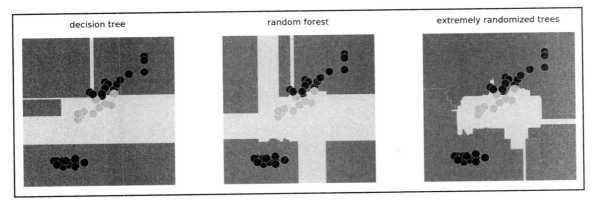

Comparing decision boundaries for various tree based classifiers

Now, the differences between the three classifiers become clearer. We see the single tree drawing by far the simplest decision boundaries, splitting the landscape using horizontal decision boundaries. The random forest is able to more clearly separate the cloud of data points in the lower left of the decision landscape. However, only extremely randomized trees were able to corner the cloud of data points towards the center of the landscape from all sides.

Now that we know about all the different variations of tree ensembles, let's move on to a real-world dataset.

Using random forests for face recognition

A popular dataset that we haven't talked much about yet is the **Olivetti face dataset**.

The Olivetti face dataset was collected in 1990 by AT&T Laboratories Cambridge. The dataset comprises facial images of 40 distinct subjects, taken at different times and under different lighting conditions. In addition, subjects varied their facial expression (open/closed eyes, smiling/not smiling) and their facial details (glasses/no glasses).

Images were then quantized to 256 grayscale levels and stored as unsigned 8-bit integers. Because there are 40 distinct subjects, the dataset comes with 40 distinct target labels. Recognizing faces thus constitutes an example of a **multiclass classification** task.

Loading the dataset

Like many other classic datasets, the Olivetti face dataset can be loaded using scikit-learn:

```
In [1]: from sklearn.datasets import fetch_olivetti_faces
   ...  dataset = fetch_olivetti_faces()
In [2]: X = dataset.data
   ...  y = dataset.target
```

Although the original images consisted of *92 x 112* pixel images, the version available through scikit-learn contains images downscaled to *64 x 64* pixels.

To get a sense of the dataset, we can plot some example images. Let's pick eight indices from the dataset in a random order:

```
In [3]: import numpy as np
   ...  np.random.seed(21)
   ...  idx_rand = np.random.randint(len(X), size=8)
```

We can plot these example images using Matplotlib, but we need to make sure we reshape the column vectors to 64 x 64 pixel images before plotting:

```
In [4]: import matplotlib.pyplot as plt
   ...  %matplotlib inline
   ...  for p, i in enumerate(idx_rand):
   ...      plt.subplot(2, 4, p + 1)
   ...  plt.imshow(X[i, :].reshape((64, 64)), cmap='gray')
   ...      plt.axis('off')
```

The preceding code produces the following picture:

Example face images from the Olivetti dataset

You can see how all the faces are taken against a dark background and are upright. The facial expression varies drastically from image to image, making this an interesting classification problem. Try not to laugh at some of them!

Preprocessing the dataset

Before we can pass the dataset to the classifier, we need to preprocess it following the best practices from `Chapter 4`, *Representing Data and Engineering Features*.

Specifically, we want to make sure that all example images have the same mean grayscale level:

```
In [5]: n_samples, n_features = X.shape[:2]
   ...    X -= X.mean(axis=0)
```

We repeat this procedure for every image to make sure the feature values of every data point (that is, a row in X) are centered around zero:

```
In [6]: X -= X.mean(axis=1).reshape(n_samples, -1)
```

The preprocessed data can be visualized using the preceding code:

```
In [7]: for p, i in enumerate(idx_rand):
   ...        plt.subplot(2, 4, p + 1)
   ...        plt.imshow(X[i, :].reshape((64, 64)), cmap='gray')
   ...        plt.axis('off')
```

This produces the following figure:

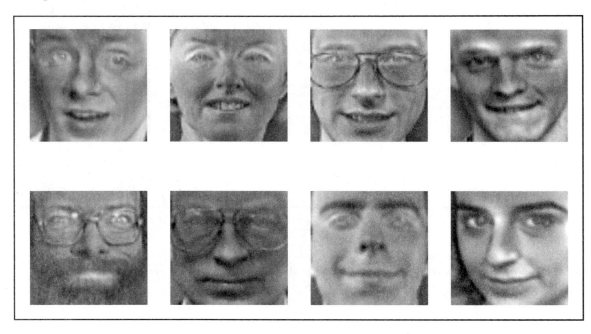

Examples of preprocessed face images from the Olivetti dataset

Training and testing the random forest

We continue to follow our best practice to split the data into training and test sets:

```
In [8]: from sklearn.model_selection import train_test_split
   ...    X_train, X_test, y_train, y_test = train_test_split(
   ...        X, y, random_state=21
   ...    )
```

Then we are ready to apply a random forest to the data:

```
In [9]: import cv2
   ...  rtree = cv2.ml.RTrees_create()
```

Here we want to create an ensemble with 50 decision trees:

```
In [10]: n_trees = 50
    ...  eps = 0.01
    ...  criteria = (cv2.TERM_CRITERIA_MAX_ITER + cv2.TERM_CRITERIA_EPS,
    ...              n_trees, eps)
    ...  rtree.setTermCriteria(criteria)
```

Because we have a large number of categories (that is, 40), we want to make sure the random forest is set up to handle them accordingly:

```
In [10]: rtree.setMaxCategories(len(np.unique(y)))
```

We can play with other optional arguments, such as the number of data points required in a node before it can be split:

```
In [11]: rtree.setMinSampleCount(2)
```

However, we might not want to limit the depth of each tree. This is again, a parameter we will have to experiment with in the end. But for now, let's set it to a large integer value, making the depth effectively unconstrained:

```
In [12]: rtree.setMaxDepth(1000)
```

Then we can fit the classifier to the training data:

```
In [13]: rtree.train(X_train, cv2.ml.ROW_SAMPLE, y_train);
```

We can check the resulting depth of the tree using the following function:

```
In [13]: rtree.getMaxDepth()
Out[13]: 25
```

This means that although we allowed the tree to go up to depth 1000, in the end only 25 layers were needed.

The evaluation of the classifier is done once again by predicting the labels first (`y_hat`) and then passing them to the `accuracy_score` function:

```
In [14]: _, y_hat = tree.predict(X_test)
In [15]: from sklearn.metrics import accuracy_score
   ...:  accuracy_score(y_test, y_hat)
Out[15]: 0.87
```

We find 87% accuracy, which turns out to be much better than with a single decision tree:

```
In [16]: from sklearn.tree import DecisionTreeClassifier
   ...:  tree = DecisionTreeClassifier(random_state=21, max_depth=25)
   ...:  tree.fit(X_train, y_train)
   ...:  tree.score(X_test, y_test)
Out[16]: 0.46999999999999997
```

Not bad! We can play with the optional parameters to see if we get better. The most important one seems to be the number of trees in the forest. We can repeat the experiment with a forest made from 100 trees:

```
In [18]: num_trees = 100
   ...:  eps = 0.01
   ...:  criteria = (cv2.TERM_CRITERIA_MAX_ITER + cv2.TERM_CRITERIA_EPS,
   ...:              num_trees, eps)
   ...:  rtree.setTermCriteria(criteria)
   ...:  rtree.train(X_train, cv2.ml.ROW_SAMPLE, y_train);
   ...:  _, y_hat = rtree.predict(X_test)
   ...:  accuracy_score(y_test, y_hat)
Out[18]: 0.91000000000000003
```

With this configuration, we get 91% accuracy!

Here we tried to improve the performance of our model through **creative trial-and-error**: We vary some of the parameters we deem important, and observe the resulting change in performance, until we find a configuration that satisfies our expectations. We will learn more sophisticated techniques for improving a model in `Chapter 11`, *Selecting the Right Model with Hyperparameter Tuning*.

Another interesting use case of decision tree ensembles is Adaptive Boosting or AdaBoost.

Implementing AdaBoost

When the trees in the forest are trees of depth 1 (also known as **decision stumps**) and we perform boosting instead of bagging, the resulting algorithm is called **AdaBoost**.

AdaBoost adjusts the dataset at each iteration by performing the following actions:

- Selecting a decision stump
- Increasing the weighting of cases that the decision stump labeled incorrectly while reducing the weighting of correctly labeled cases

This iterative weight adjustment causes each new classifier in the ensemble to prioritize training the incorrectly labeled cases. As a result, the model adjusts by targeting highly-weighted data points.

Eventually, the stumps are combined to form a final classifier.

Implementing AdaBoost in OpenCV

Although OpenCV provides a very efficient implementation of AdaBoost, it is hidden under the **Haar cascade** classifier. Haar cascade classifiers are a very popular tool for face detection, which we can illustrate through the example of the Lena image:

```
In [1]: img_bgr = cv2.imread('data/lena.jpg', cv2.IMREAD_COLOR)
   ...: img_gray = cv2.cvtColor(img_bgr, cv2.COLOR_BGR2GRAY)
```

After loading the image in both color and grayscale, we load a pretrained Haar cascade:

```
In [2]: import cv2
   ...: filename = 'data/haarcascade_frontalface_default.xml'
   ...: face_cascade = cv2.CascadeClassifier(filename)
```

The classifier will then detect faces present in the image using the following function call:

```
In [3]: faces = face_cascade.detectMultiScale(img_gray, 1.1, 5)
```

Note that the algorithm operates only on grayscale images. That's why we saved two pictures of Lena, one to which we can apply the classifier (`img_gray`), and one on which we can draw the resulting bounding box (`img_bgr`):

```
In [4]: color = (255, 0, 0)
   ...: thickness = 2
   ...: for (x, y, w, h) in faces:
   ...:     cv2.rectangle(img_bgr, (x, y), (x + w, y + h),
   ...:                   color, thickness)
```

Then we can plot the image using the following code:

```
In [5]: import matplotlib.pyplot as plt
   ...: %matplotlib inline
   ...: plt.imshow(cv2.cvtColor(img_bgr, cv2.COLOR_BGR2RGB));
```

This results in the following image, with the location of the face indicated by a blue bounding box:

Detecting faces in the Lena image

Obviously, this picture contains only a single face. However, the preceding code will work even on images where multiple faces could be detected. Try it out!

Implementing AdaBoost in scikit-learn

In scikit-learn, AdaBoost is just another ensemble estimator. We can create an ensemble from 100 decision stumps as follows:

```
In [6]: from sklearn.ensemble import AdaBoostClassifier
   ...     ada = AdaBoostClassifier(n_estimators=100,
   ...                              random_state=456)
```

We can load the breast cancer set once more and split it 75-25:

```
In [7]: from sklearn.datasets import load_breast_cancer
   ...     cancer = load_breast_cancer()
   ...     X = cancer.data
   ...     y = cancer.target
In [8]: from sklearn.model_selection import train_test_split
   ...     X_train, X_test, y_train, y_test = train_test_split(
   ...         X, y, random_state=456
   ...     )
```

Then `fit` and `score` AdaBoost using the familiar procedure:

```
In [9]: ada.fit(X_train, y_train)
   ...     ada.score(X_test, y_test)
Out[9]: 0.97902097902097907
```

The result is remarkable, 97.9% accuracy!

We might want to compare this result to a random forest. However, to be fair, we should make the trees in the forest all decision stumps. Then we will know the difference between bagging and boosting:

```
In [10]: from sklearn.ensemble import RandomForestClassifier
   ...     forest = RandomForestClassifier(n_estimators=100,
   ...                                     max_depth=1,
   ...                                     random_state=456)
   ...     forest.fit(X_train, y_train)
   ...     forest.score(X_test, y_test)
Out[10]: 0.93706293706293708
```

Of course, if we let the trees be as deep as needed, we might get a better score:

```
In [11]: forest = RandomForestClassifier(n_estimators=100,
   ...                                   random_state=456)
   ...     forest.fit(X_train, y_train)
   ...     forest.score(X_test, y_test)
Out[11]: 0.98601398601398604
```

As a last step in this chapter, let's talk about how to combine different types of models into an ensemble.

Combining different models into a voting classifier

So far, we saw how to combine different instances of the same classifier or regressor into an ensemble. In this chapter, we are going to take this idea a step further and combine conceptually different classifiers into what is known as a **voting classifier**.

The idea behind voting classifiers is that the individual learners in the ensemble don't necessarily need to be of the same type. After all, no matter how the individual classifiers arrived at their prediction, in the end, we are going to apply a decision rule that integrates all the votes of the individual classifiers. This is also known as a **voting scheme**.

Understanding different voting schemes

Two different voting schemes are common among voting classifiers:

- In **hard voting** (also known as **majority voting**), every individual classifier votes for a class, and the majority wins. In statistical terms, the predicted target label of the ensemble is the mode of the distribution of individually predicted labels.
- In **soft voting**, every individual classifier provides a probability value that a specific data point belongs to a particular target class. The predictions are weighted by the classifier's importance and summed up. Then the target label with the greatest sum of weighted probabilities wins the vote.

For example, let's assume we have three different classifiers in the ensemble that perform a binary classification task. Under the hard voting scheme, every classifier would predict a target label for a particular data point:

classifier	predicted target label
classifier #1	class 1
classifier #2	class 1
classifier #3	class 0

The voting classifier would then tally up the votes and go with the majority. In this case, the ensemble classifier would predict class 1.

In a soft voting scheme, the math is slightly more involved. In soft voting, every classifier is assigned a weight coefficient, which stands for the importance of the classifier in the voting procedure. For the sake of simplicity, let's assume all three classifiers have the same weight: $w_1 = w_2 = w_3 = 1$.

The three classifiers would then go on to predict the probability of a particular data point belonging to each of the available class labels:

classifier	class 0	class 1
classifier #1	$0.3\,w_1$	$0.7\,w_1$
classifier #2	$0.5\,w_2$	$0.5\,w_2$
classifier #3	$0.4\,w_3$	$0.6\,w_3$

In this example, classifier #1 is 70% sure we're looking at an example of class 1. Classifier #2 is fifty-fifty, and classifier #3 tends to agree with classifier #1. Every probability score gets combined with the weight coefficient of the classifier.

The voting classifier would then compute the weighted average for each class label:

- For class 0, we get a weighted average of $(0.3\,w_1 + 0.5\,w_2 + 0.4\,w_3) / 3 = 0.4$
- For class 1, we get a weighted average of $(0.7\,w_1 + 0.5\,w_2 + 0.6\,w_3) / 3 = 0.6$

Because the weighted average for class 1 is higher than for class 0, the ensemble classifier would go on to predict class 1.

Implementing a voting classifier

Let's look at a simple example of a voting classifier that combines three different algorithms:

- A logistic regression classifier from Chapter 3, *First Steps in Supervised Learning*
- A Gaussian naive Bayes classifier from Chapter 7, *Implementing a Spam Filter with Bayesian Learning*
- A random forest classifier from this chapter

Combining Different Algorithms into an Ensemble

We can combine these three algorithms into a voting classifier and apply it to the breast cancer dataset with the following steps:

1. Load the dataset, and split it into training and test sets:

   ```
   In [1]: from sklearn.datasets import load_breast_cancer
      ...: cancer = load_breast_cancer()
      ...: X = cancer.data
      ...: y = cancer.target
   In [2]: from sklearn.model_selection import train_test_split
      ...: X_train, X_test, y_train, y_test = train_test_split(
      ...:     X, y, random_state=13
      ...: )
   ```

2. Instantiate the individual classifiers:

   ```
   In [3]: from sklearn.linear_model import LogisticRegression
      ...: model1 = LogisticRegression(random_state=13)
   In [4]: from sklearn.naive_bayes import GaussianNB
      ...: model2 = GaussianNB()
   In [5]: from sklearn.ensemble import RandomForestClassifier
      ...: model3 = RandomForestClassifier(random_state=13)
   ```

3. Assign the individual classifiers to the voting ensemble. Here, we need to pass a list of tuples (`estimators`), where every tuple consists of the name of the classifier (a string of our choosing) and the model object. The voting scheme can be either `voting='hard'` or `voting='soft'`:

   ```
   In [6]: from sklearn.ensemble import VotingClassifier
      ...: vote = VotingClassifier(estimators=[('lr', model1),
      ...:                                     ('gnb', model2),
      ...:                                     ('rfc', model3)],
      ...:                         voting='hard')
   ```

4. Fit the ensemble to the training data and score it on the test data:

   ```
   In [7]: vote.fit(X_train, y_train)
      ...: vote.score(X_test, y_test)
   Out[7]: 0.95104895104895104
   ```

In order to convince us that 95.1% is a great accuracy score, we can compare the ensemble's performance to the theoretical performance of each individual classifier. We do this by fitting the individual classifiers to the data. Then we will see that the logistic regression model achieves 94.4% accuracy on its own:

```
In [8]: model1.fit(X_train, y_train)
   ...  model1.score(X_test, y_test)
Out[8]: 0.94405594405594406
```

Similarly, the naive Bayes classifier achieves 93.0% accuracy:

```
In [9]: model2.fit(X_train, y_train)
   ...  model2.score(X_test, y_test)
Out[9]: 0.93006993006993011
```

Last but not least, the random forest classifier also achieved 94.4% accuracy:

```
In [10]: model3.fit(X_train, y_train)
   ...   model3.score(X_test, y_test)
Out[10]: 0.94405594405594406
```

All in all, we were just able to gain a good percent in performance by combining three unrelated classifiers into an ensemble. Each of these classifiers might have made different mistakes on the training set, but that's OK because on average, we need just two out of three classifiers to be correct.

Summary

In this chapter, we talked about how to improve various classifiers by combining them into an ensemble. We discussed how to average the predictions of different classifiers using bagging and how to have different classifiers correct each other's mistakes using boosting. A lot of time was spent discussing all possible ways to combine decision trees, be it decision stumps (AdaBoost), a bag of full of trees, random forests, or extremely randomized trees. Finally, we learned how to combine even different types of classifiers in an ensemble by building a voting classifier.

In the next chapter, we will talk more about how to compare the results of different classifiers by diving into the world of model selection and hyperparameter tuning.

11
Selecting the Right Model with Hyperparameter Tuning

Now that we have visited a wide variety of machine learning algorithms, I am sure you have realized that most of them come with a great number of settings to choose from. These settings or tuning knobs, the so-called **hyperparameters**, help us control the behavior of the algorithm when we try to maximize performance.

For example, we might want to choose the depth or split criterion in a decision tree or tune the number of neurons in a neural network. Finding the values of important parameters of a model is a tricky task but necessary for almost all models and datasets.

In this chapter, we will thus dive deeper into **model evaluation** and **hyperparameter tuning**. Assume that we have two different models that might apply to our task. How can we know which one is better? Answering this question often involves repeatedly fitting different versions of our model to different subsets of the data, such as in **cross-validation** and **bootstrapping**. In combination with different scoring functions, we can obtain reliable estimates of the generalization performance of our models.

But what if two different models give similar results? Can we be sure that the two models are equivalent, or is it possible that one of them just got lucky? How can we know whether one of them is significantly better than the other? Answering these questions will lead us to discussing some useful statistical tests such as **Students t-test** and **McNemar's test**.

As we will get familiar with these techniques, we will also want to answer the following questions:

- What's the best strategy to tweak the hyperparameters of a model?
- How can we compare the performance of different models in a fair way?
- How do we select the right machine learning tool for the task at hand?

Evaluating a model

Model evaluation strategies come in many different forms and shapes. In the following sections, we will, therefore, highlight three of the most commonly used techniques to compare models against each other:

- k-fold cross-validation
- bootstrapping
- McNemar's test

In principle, model evaluation is simple: after training a model on some data, we can estimate its effectiveness by comparing model predictions to some ground truth values. We learned early on that we should split the data into a training and a test set, and we tried to follow this instruction whenever possible. But why exactly did we do that again?

Evaluating a model the wrong way

The reason we never evaluate a model on the training set is that, in principle, any dataset can be learned if we throw a strong enough model at it.

A quick demonstration of this can be given with help of the Iris dataset, which we talked about extensively in Chapter 3, *First Steps in Supervised Learning*. There the goal was to classify species of Iris flowers based on their physical dimensions. We can load the Iris dataset using scikit-learn:

```
In [1]: from sklearn.datasets import load_iris
   ...     iris = load_iris()
```

An innocent approach to this problem would be to store all data points in matrix X and all class labels in the vector y:

```
In [2]: import numpy as np
   ...     X = iris.data.astype(np.float32)
   ...     y = iris.target
```

Next, we choose a model and its hyperparameters. For example, let's use the k-NN algorithm from Chapter 3, *First Steps in Supervised Learning*, which provides only a single hyperparameter: the number of neighbors, k. With $k=1$, we get a very simple model that classifies the label of an unknown point as belonging to the same class as its closest neighbor.

In OpenCV, k-NN instantiates as follows:

```
In [3]: import cv2
    ...     knn = cv2.ml.KNearest_create()
    ...     knn.setDefaultK(1)
```

Then we train the model and use it to predict labels for the data that we already know:

```
In [4]: knn.train(X, cv2.ml.ROW_SAMPLE, y)
    ...     _, y_hat = knn.predict(X)
```

Finally, we compute the fraction of correctly labeled points:

```
In [5]: from sklearn.metrics import accuracy_score
    ...     accuracy_score(y, y_hat)
Out[5]: 1.0
```

We see an accuracy score of 1.0, which indicates that 100% of points were correctly labeled by our model!

If a model gets 100% accuracy on the training set, we say the model **memorized** the data.

But is this truly measuring the expected accuracy? Have we really come up with a model that we expect to be correct 100% of the time?

As you may have gathered, the answer is no. This example shows that even a simple algorithm is capable of memorizing a real-world dataset. Imagine how easy this task would have been for a deep neural network! Usually, the more parameters a model has, the more powerful it is. We will come back to this shortly.

Evaluating a model in the right way

A better sense of a model's performance can be found using what's known as a test set, but you already knew this. When presented with data held out from the training procedure, we can check whether a model has learned some dependencies in the data that hold across the board or whether it just memorized the training set.

We can split the data into training and test sets using the familiar `train_test_split` from scikit-learn's `model_selection` module:

```
In [6]: from sklearn.model_selection import train_test_split
```

But how do we choose the right train-test ratio? Is there even such a thing as a right ratio? Or is this considered another hyperparameter of the model?

There are two competing concerns here:

- If our training set is too small, our model might not be able to extract the relevant data dependencies. As a result, our model performance might differ significantly from run to run, that is, if we repeat our experiment multiple times with different random number seeds. As an extreme example, consider a training set with a single data point from the Iris dataset. In this case, there would be no way for the model to even learn that there are multiple species in the dataset!
- If our test set is too small, our performance metric might differ significantly from run to run. As a result, we would have to rerun our experiment multiple times to get an idea of how well our model does on average. As an extreme example, consider a test set with a single data point. Since there are three different classes in the Iris dataset, we might get either 0, 33%, 66%, or 100% correct.

A good starting point is usually a 80-20 training-test split. However, it all depends on the amount of data available. For relatively small datasets, a 50-50 split might be more suitable:

```
In [7]: X_train, X_test, y_train, y_test = train_test_split(
   ...         X, y, random_state=37, train_size=0.5
   ...     )
```

Then we retrain the preceding model on the training set:

```
In [8]: knn = cv2.ml.KNearest_create()
   ...  knn.setDefaultK(1)
   ...  knn.train(X_train, cv2.ml.ROW_SAMPLE, y_train)
```

When we test the model on the test set, we suddenly get a different result:

```
In [9]: _, y_test_hat = knn.predict(X_test)
   ...  accuracy_score(y_test, y_test_hat)
Out[9]: 0.96666666666666667
```

We see a more reasonable result here, although 97% accuracy is still a formidable result. But is this the best possible result—and how can we know for sure?

To answer this question is, we have to dig a little deeper.

Selecting the best model

When a model is under-performing, it is often not clear how to make it better. Throughout this book, I have declared several decisions a secret of the trade, for example, how to select the number of layers in a neural network. Even worse, the answer is often counter intuitive! For example, adding another layer to the network might make the results worse, and adding more training data might not change performance at all.

You can see why these issues are some of the most important aspects of machine learning. At the end of the day, the ability to determine what steps will or will not improve our model is what separates the successful machine learning practitioner from all others.

Let's have a look at a specific example. Remember Chapter 5, *Using Decision Trees to Make a Medical Diagnosis*, where we used decision trees in a regression task? We were fitting two different trees to a sin function—one with depth 2 and one with depth 5. As a reminder, the regression result looked like this:

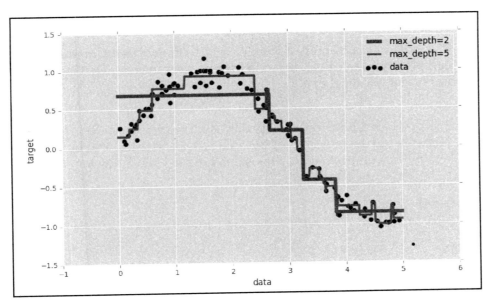

Approximating a sin function using decision trees

It should be clear that neither of these fits are particularly good. However, the two decision trees fail in two different ways!

The decision tree with depth 2 (thick line in the preceding figure) attempts to fit four straight lines through the data. Because the data is intrinsically more complicated than a few straight lines, this model fails. We could train it as much as we wanted, on as many training samples as we could generate—it would never be able to describe this dataset well. Such a model is said to **underfit** the data. In other words, the model does not have enough flexibility to account for all the features in the data. Another way of saying this is that the model has **high bias**.

The other decision tree (thin line, depth 5) makes a different mistake. This model has enough flexibility to nearly perfectly account for the fine structures in the data. However, at some points, the model seems to follow the particular pattern of the noise; we added to the sin function rather than the sin function itself. You can see that on the right-hand side of the graph, where the blue curve (thin line) would jitter a lot. Such a model is said to **overfit** the data. In other words, the model has so much flexibility that it ends up accounting for random errors in the data. Another way of saying this is that model has **high variance**.

To give you a long story short—here's the secret sauce: Fundamentally, selecting the right model comes down to finding a sweet spot in the trade-off between bias and variance.

The amount of flexibility a model has (also known as the **model complexity**) is mostly dictated by its hyperparameters. That is why it is so important to tune them!

Let's return to the *k*-NN algorithm and the Iris dataset. If we repeated the procedure of fitting the model to the Iris data for all possible values of *k* and calculated both training and test scores, we would expect the result to look something like the following:

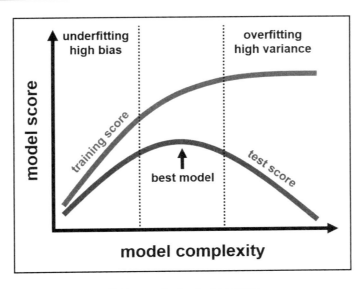

Model score as a function of model complexity

If there is one thing I would want you to remember from this chapter, it would be this diagram. Let's unpack it.

The diagram describes the model score (either training or test score) as a function of model complexity. As mentioned in the preceding diagram, the model complexity of a neural network roughly grows with the number of neurons in the network. In the case of *k*-NN, the opposite logic applies—the larger the value for *k*, the smoother the decision boundary, and thus, the lower the complexity. In other words, *k*-NN with *k*=1 would be all the way to the right in the preceding diagram, where the training score is perfect. No wonder we got 100% accuracy on the training set!

From the preceding diagram, we can gather that there are three regimes in the model complexity landscape:

- For very low model complexity (a high-bias model), the training data is underfit. In this regime, the model achieves only low scores on both the training and test set, no matter for how long we trained it.
- For very high model complexity (a high-variance model), the training data is overfit, which means that the model predicts the training data very well but fails for any previously unseen data. In this regime, the model has started to learn intricacies or peculiarities that only appear in the training data. Since these peculiarities do not apply to unseen data, the training score gets lower and lower.

- For some intermediate value, the test score is maximal. It is this intermediate regime, where the test score is maximal, that we are trying to find. This is the sweet spot in the trade-off between bias and variance!

This means that we can find the best algorithm for the task at hand by mapping out the model complexity landscape. Specifically, we can use the following indicators to know which regime we are currently in:

- If both training and test scores are below our expectations, we are probably in the leftmost regime in the preceding diagram, where the model is underfitting the data. In this case, a good idea might be to increase the model complexity and try again.
- If the training score is much higher than the test score, we are probably in the rightmost regime in the preceding diagram, where model is overfitting the data. In this case, a good idea might be to decrease the model complexity and try again.

Although this procedure works in general, there are more sophisticated strategies for model evaluation that proved to be more thorough than a simple train-test split, which we will talk about in the following sections.

Understanding cross-validation

Cross-validation is a method of evaluating the generalization performance of a model that is generally more stable and thorough than splitting the dataset into training and test sets.

The most commonly used version of cross-validation is *k*-**fold cross-validation**, where *k* is a number specified by the user (usually five or ten). Here, the dataset is partitioned into *k* parts of more or less equal size, called **folds**. For a dataset that contains *N* data points, each fold should thus have approximately *N* / *k* samples. Then a series of models is trained on the data, using *k* - *1* folds for training and one remaining fold for testing. The procedure is repeated for *k* iterations, each time choosing a different fold for testing, until every fold has served as a test set once

An example of two-fold cross-validation is shown in the following figure:

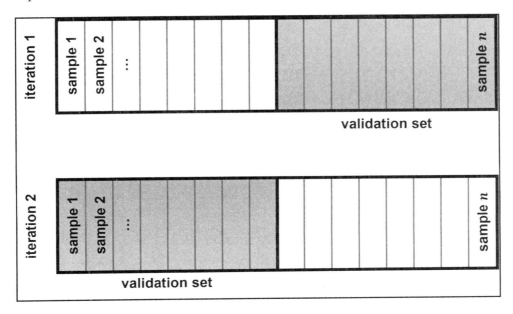

Two-fold cross-validation

At first glance, this might look like our regular procedure of splitting the data into a training and test set. One difference is that cross-validation is run twice in this case: once where the model is trained on what we usually call the training set (and tested on the test set), and once where the model is trained on what we usually call the test set (and tested on the training set).

Analogously, in a cross-validation procedure with four folds, we would start by using the first three folds for training and the fourth fold for testing (iteration 1). In iteration 2, we would choose the third fold for testing (called **validation set** shown in the following figure) and the remaining folds (folds 1, 2, and 4) for testing, and so on:

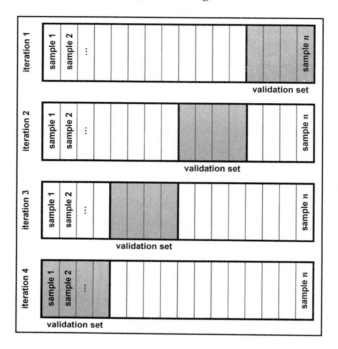

Four-fold cross-validation

During each of these iterations, we compute the accuracy on the validation set. At the end of the day, we take the average of all accuracy scores to get a final mean score of the model. This will give us a much more robust estimate of the generalization performance of our model because we repeated the train-test procedure multiple times on multiple random splits of the data.

 Note that the purpose of cross-validation is not to build a final model, that is, we don't use these four instances of our trained model to do any real prediction. The purpose of cross-validation is only to evaluate how well a given algorithm will generalize when trained on a specific dataset.

Having multiple splits of the data also provides some information about how sensitive our model is to the selection of the training dataset. Since we are dealing with random data splits, we have to consider the possibility that our procedure is subject to luck:

- We might get **lucky** in one iteration so that for some reason, all data points in the validation set are really easy to classify. Then the reported accuracy would be unreasonably high.
- We might get **unlucky** in one iteration so that the data points in the validation set are practically impossible to classify correctly. Then the reported accuracy would be unreasonably low.

By repeating the procedure for *k* folds, we make sure that the observed result does not strongly depend on the exact train-test split. As a bonus, we can get an idea of how the model might perform in the worst case and best case scenarios when applied to new data by looking at the differences in accuracy from one fold to the next.

Another benefit of cross-validation over a single train-test split is that we use our data more effectively. Usually, the more data we can use to fit the model, the more accurate the model will end up being. However, if we did a 80-20 train-test split using the train_test_split function, we would have to reserve 20% of all data points for testing. In five-fold cross-validation, we might set 20% of the data points for validation, but by cycling through the different iterations, we eventually end up using every data point for training. Even better—in ten-fold cross-validation, we can use 90% of the data for training!

Manually implementing cross-validation in OpenCV

The easiest way to perform cross-validation in OpenCV is to do the data splits by hand.

For example, in order to implement two-fold cross-validation, we would follow the following procedure:

1. Load the dataset:

   ```
   In [1]: from sklearn.datasets import load_iris
      ...: import numpy as np
      ...: iris = load_iris()
      ...: X = iris.data.astype(np.float32)
      ...: y = iris.target
   ```

2. Split the data into two equally sized parts:

   ```
   In [2]: from sklearn.model_selection import model_selection
      ...: X_fold1, X_fold2, y_fold1, y_fold2 = train_test_split(
      ...:     X, y, random_state=37, train_size=0.5
      ...: )
   ```

3. Instantiate the classifier:

```
In [3]: import cv2
   ...     knn = cv2.ml.KNearest_create()
   ...     knn.setDefaultK(1)
```

4. Train the classifier on the first fold, then predict the labels of the second fold:

```
In [4]: knn.train(X_fold1, cv2.ml.ROW_SAMPLE, y_fold1)
   ...     _, y_hat_fold2 = knn.predict(X_fold2)
```

5. Train the classifier on the second fold, then predict the labels of the first fold:

```
In [5]: knn.train(X_fold2, cv2.ml.ROW_SAMPLE, y_fold2)
   ...     _, y_hat_fold1 = knn.predict(X_fold1)
```

6. Compute accuracy scores for both folds:

```
In [6]: from sklearn.metrics import accuracy_score
   ...     accuracy_score(y_fold1, y_hat_fold1)
Out[6]: 0.92000000000000004
In [7]: accuracy_score(y_fold2, y_hat_fold2)
Out[7]: 0.88
```

This procedure will yield two accuracy scores, one for the first fold (92% accuracy), and one for the second fold (88% accuracy). On average, our classifier thus achieved 90% accuracy on unseen data.

Using scikit-learn for k-fold cross-validation

In scikit-learn, cross-validation can be performed in three steps:

1. Load the dataset. Since we already did this earlier, we don't have to do it again.
2. Instantiate the classifier:

```
In [8]: from sklearn.neighbors import KNeighborsClassifier
   ...     model = KNeighborsClassifier(n_neighbors=1)
```

3. Perform cross-validation with the `cross_val_score` function. This function takes as input a model, the full dataset (X), the target labels (y) and an integer value for the number of folds (cv). It is not necessary to split the data by hand—the function will do that automatically depending on the number of folds. After the cross-validation is completed, the function returns the test scores:

```
In [9]: from sklearn.model_selection import cross_val_score
   ...: scores = cross_val_score(model, X, y, cv=5)
   ...: scores
Out[9]: array([ 0.96666667, 0.96666667, 0.93333333, 0.93333333,
        1. ])
```

In order to get a sense how the model did on average, we can look at the mean and standard deviation of the five scores:

```
In [10]: scores.mean(), scores.std()
Out[10]: (0.95999999999999996, 0.024944382578492935)
```

With five folds, we have a much better idea about how robust the classifier is on average. We see that k-NN with k=1 achieves on average 96% accuracy, and this value fluctuates from run to run with a standard deviation of roughly 2.5%.

Implementing leave-one-out cross-validation

Another popular way to implement cross-validation is to choose the number of folds equal to the number of data points in the dataset. In other words, if there are N data points, we set k=N. This means that we will end up having to do N iterations of cross-validation, but in every iteration, the training set will consist of only a single data point. The advantage of this procedure is that we get to use all-but-one data point for training. Hence, this procedure is also known as **leave-one-out** cross-validation.

In scikit-learn, this functionality is provided by the LeaveOneOut method from the model_selection module:

```
In [11]: from sklearn.model_selection import LeaveOneOut
```

This object can be passed directly to the cross_val_score function in the following way:

```
In [12]: scores = cross_val_score(model, X, y, cv=LeaveOneOut())
```

Because every test set now contains a single data point, we would expect the scorer to return 150 values—one for each data point in the dataset. Each of these points we could get either right or wrong. Thus, we expect `scores` to be a list of ones (1) and zeros (0), which corresponds to correct and incorrect classifications, respectively:

```
In [13]: scores
Out[13]: array([ 1., 1., 1., 1., 1., 1., 1., 1., 1., 1., 1., 1., 1.,
                1., 1., 1., 1., 1., 1., 1., 1., 1., 1., 1., 1., 1.,
                1., 1., 1., 1., 1., 1., 1., 1., 1., 1., 1., 1., 1.,
                1., 1., 1., 1., 1., 1., 1., 1., 1., 1., 1., 1., 1.,
                1., 1., 1., 1., 1., 1., 1., 1., 1., 1., 1., 1., 1.,
                1., 1., 1., 1., 1., 0., 1., 0., 1., 1., 1., 1., 1.,
                1., 1., 1., 1., 1., 0., 1., 1., 1., 1., 1., 1., 1.,
                1., 1., 1., 1., 1., 1., 1., 1., 1., 1., 1., 1., 1.,
                1., 1., 0., 1., 1., 1., 1., 1., 1., 1., 1., 1., 1.,
                1., 1., 0., 1., 1., 1., 1., 1., 1., 1., 1., 1., 1.,
                1., 1., 1., 0., 1., 1., 1., 1., 1., 1., 1., 1., 1.,
                1., 1., 1., 1., 1., 1., 1.])
```

If we want to know the average performance of the classifier, we would still compute the mean and standard deviation of the scores:

```
In [14]: scores.mean(), scores.std()
Out[14]: (0.95999999999999996, 0.19595917942265423)
```

We can see this scoring scheme returns very similar results to five-fold cross-validation.

 You can learn more about other useful cross-validation procedures at `http://scikit-learn.org/stable/modules/cross_validation.html`.

Estimating robustness using bootstrapping

An alternative procedure to *k*-fold cross-validation is **bootstrapping**.

Instead of splitting the data into folds, bootstrapping builds a training set by drawing samples randomly from the dataset. Typically, a bootstrap is formed by drawing samples with replacement. Imagine putting all of the data points into a bag and then drawing randomly from the bag. After drawing a sample, we would put it back in the bag. This allows for some samples to show up multiple times in the training set, which is something cross-validation does not allow.

The classifier is then tested on all samples that are not part of the bootstrap (the so-called **out-of-bag** examples), and the procedure is repeated a large number of times (say, 10,000 times). Thus, we get a distribution of the model's score that allows us to estimate the robustness of the model.

Manually implementing bootstrapping in OpenCV

Bootstrapping can be implemented with the following procedure:

1. Load the dataset. Since we already did this earlier, we don't have to do it again.
2. Instantiate the classifier:

   ```
   In [15]: knn = cv2.ml.KNearest_create()
   ...      knn.setDefaultK(1)
   ```

3. From our dataset with N samples, randomly choose N samples with replacement to form a bootstrap. This can be done most easily with the `choice` function from NumPy's `random` module. We tell the function to draw `len(X)` samples in the range `[0, len(X)-1]` with replacement (`replace=True`). The function then returns a list of indices, from which we form our bootstrap:

   ```
   In [16]: idx_boot = np.random.choice(len(X), size=len(X),
   ...                                  replace=True)
   ...      X_boot = X[idx_boot, :]
   ...      y_boot = y[idx_boot]
   ```

4. Put all samples that do not show in the bootstrap in the out-of-bag set:

   ```
   In [17]: idx_oob = np.array([x not in idx_boot
   ...                          for x in np.arange(len(X))],dtype=np.bool)
   ...      X_oob = X[idx_oob, :]
   ...      y_oob = y[idx_oob]
   ```

5. Train the classifier on the bootstrap samples:

   ```
   In [18]: knn.train(X_train, cv2.ml.ROW_SAMPLE, y_boot)
   Out[18]: True
   ```

6. Test the classifier on the out-of-bag samples:

   ```
   In [19]: _, y_hat = knn.predict(X_oob)
   ...      accuracy_score(y_oob, y_hat)
   Out[19]: 0.9285714285714286
   ```

7. Steps 3-6 form one it

8. Iteration of the bootstrap. Repeat these steps up to 10,000 times to get 10,000 accuracy scores, then average the scores to get an idea of the classifier's mean performance.

For our convenience, we can build a function from steps 3-6 so that it is easy to run the procedure for some `n_iter` number of times. We also pass a model (our k-NN classifier, `model`), the feature matrix (X), and the vector with all class labels (y):

```
In [20]: def yield_bootstrap(model, X, y, n_iter=10000):
    ...      for _ in range(n_iter):
```

The steps within the for loop are essentially steps 3-6 from the code mentioned earlier. This involved training the classifier on the bootstrap and testing it on the out-of-bag examples:

```
    ...         # train the classifier on bootstrap
    ...         idx_boot = np.random.choice(len(X), size=len(X),
                                            replace=True)
    ...         X_boot = X[idx_boot, :]
    ...         y_boot = y[idx_boot]
    ...         knn.train(X_boot, cv2.ml.ROW_SAMPLE, y_boot)
    ...
    ...         # test classifier on out-of-bag examples
    ...         idx_oob = np.array([x not in idx_boot
    ...                             for x in np.arange(len(X))],
    ...                            dtype=np.bool)
    ...         X_oob = X[idx_oob, :]
    ...         y_oob = y[idx_oob]
    ...         _, y_hat = knn.predict(X_oob)
```

Then we need to return the accuracy score. You might expect a `return` statement here. However, a more elegant way is to use the `yield` statement, which turns the function automatically into a generator. This means we don't have to initialize an empty list (`acc = []`) and then append the new accuracy score at each iteration (`acc.append(accuracy_score(...))`). The bookkeeping is done automatically:

```
    ...         yield accuracy_score(y_oob, y_hat)
```

To make sure we all get the same result, let's fix the seed of the random number generator:

```
In [21]: np.random.seed(42)
```

Now, let's run the procedure for `n_iter=10` times by converting the function output to a list:

```
In [22]: list(yield_bootstrap(knn, X, y, n_iter=10))
Out[22]: [0.98333333333333328,
         0.93650793650793651,
         0.92452830188679247,
         0.92307692307692313,
         0.94545454545454544,
         0.94736842105263153,
         0.98148148148148151,
         0.96078431372549022,
         0.93220338983050843,
         0.96610169491525422]
```

As you can see, for this small sample we get accuracy scores anywhere between 92% and 98%. To get a more reliable estimate of the model's performance, we repeat the procedure 1,000 times and calculate both mean and standard deviation of the resulting scores:

```
In [23]: acc = list(yield_bootstrap(knn, X, y, n_iter=1000))
   ...:  np.mean(acc), np.std(acc)
Out[23]: (0.95524155136419198, 0.022040380995646654)
```

You are always welcome to increase the number of repetitions. But once `n_iter` is large enough, the procedure should be robust to the randomness of the sampling procedure. In this case, we do not expect to see any more changes to the distribution of score values as we keep increasing `n_iter` to, for example, 10,000 iterations:

```
In [24]: acc = list(yield_bootstrap(knn, X, y, n_iter=10000))
   ...:  np.mean(acc), np.std(acc)
Out[24]: (0.95501528733009422, 0.021778543317079499)
```

Typically, the scores obtained with bootstrapping would be used in a **statistical test** to assess the **significance** of our result. Let's have a look at how that is done.

Assessing the significance of our results

Assume for a moment that we implemented the cross-validation procedure for two versions of our *k*-NN classifier. The resulting test scores are— 92.34% for Model A, and 92.73% for Model B. How do we know which model is better?

Following our logic introduced here, we might argue for Model B because it has a better test score. But what if the two models are not significantly different? These could have two underlying causes, which are both a consequence of the randomness of our testing procedure:

- For all we know, Model B just got lucky. Perhaps we chose a really low k for our cross-validation procedure. Perhaps Model B ended up with a beneficial train-test split so that the model had no problem classifying the data. After all, we didn't run tens of thousands of iterations like in bootstrapping to make sure the result holds in general.
- Variability in the test scores is so high that we cannot know for sure whether the two results are essentially the same. This could be the case even over 10,000 iterations in bootstrapping! If the testing procedure is inherently noisy, the resulting test scores will be noisy, too.

So how can we know for sure that the two models are different?

Implementing Student's t-test

One of the most famous statistical tests is **Student's *t*-test**. You might have heard of it before: it allows us to determine whether two sets of data are significantly different from one another. This was a really important test for William Sealy Gosset, the inventor of the test, who worked at the Guinness brewery and wanted to know whether two batches of stout differed in quality.

 Note that Student here is capitalized. Although Gosset wasn't allowed to publish his test due to company policy, he did so anyway under his pen name Student.

In practice, the *t*-test allows us to determine whether two data samples come from underlying distributions with the same mean or **expected value**.

For our purposes, this means that we can use the *t*-test to determine whether the test scores of two independent classifiers have the same mean value. We start by hypothesizing that the two sets of test scores are identical. We call this the **null hypothesis** because this is the hypothesis we want to nullify, that is, we are looking for evidence to **reject** the hypothesis because we want to ensure that one classifier is significantly better than the other.

We accept or reject a null hypothesis based on a parameter known as the *p*-**value** that the *t*-test returns. The *p*-value takes on values between 0 and 1. A *p*-value of 0.05 would mean that the null hypothesis is right only 5 out of 100 times. A small *p*-value thus indicates strong evidence that the hypothesis can be safely rejected. It is customary to use *p*=0.05 as a cut-off value below which we reject the null hypothesis.

If this is all too confusing, think of it this way: when we run a t-test for the purpose of comparing classifier test scores, we are looking to obtain a small *p*-value because that means that the two classifiers give significantly different results.

We can implement Student's *t*-test with SciPy's `ttest_ind` function from the `stats` module:

```
In [25]: from scipy.stats import ttest_ind
```

Let's start with a simple example. Assume we ran five-fold cross-validation on two classifiers and obtained the following scores:

```
In [26]: scores_a = [1, 1, 1, 1, 1]
    ...  scores_b = [0, 0, 0, 0, 0]
```

This means that Model A achieved 100% accuracy in all five folds, whereas Model B got 0% accuracy. In this case, it is clear that the two results are significantly different. If we run the *t*-test on this data, we should thus find a really small *p*-value:

```
In [27]: ttest_ind(scores_a, scores_b)
Out[27]: Ttest_indResult(statistic=inf, pvalue=0.0)
```

And we do! We actually get the smallest possible *p*-value, *p*=0.0.

On the other hand, what if the two classifiers got exactly the same numbers, except during different folds. In this case, we would expect the two classifiers to be equivalent, which is indicated by a really large *p*-value:

```
In [28]: scores_a = [0.9, 0.9, 0.9, 0.8, 0.8]
    ...  scores_b = [0.8, 0.8, 0.9, 0.9, 0.9]
    ...  ttest_ind(scores_a, scores_b)
Out[28]: Ttest_indResult(statistic=0.0, pvalue=1.0)
```

Analogous to the aforementioned, we get the largest possible *p*-value, *p*=1.0.

Selecting the Right Model with Hyperparameter Tuning

To see what happens in a more realistic example, let's return to our *k*-NN classifier from earlier example. Using the test scores obtained from the ten-fold cross-validation procedure, we can compare two different *k*-NN classifiers with the following procedure:

1. Obtain a set of test scores for Model A. We choose Model A to be the *k*-NN classifier from earlier ($k=1$):

   ```
   In [29]: k1 = KNeighborsClassifier(n_neighbors=1)
    ...     scores_k1 = cross_val_score(k1, X, y, cv=10)
    ...     np.mean(scores_k1), np.std(scores_k1)
   Out[29]: (0.95999999999999996, 0.053333333333333323)
   ```

2. Obtain a set of test scores for Model B. Let's choose Model B to be a k-NN classifier with $k=3$:

   ```
   In [30]: k3 = KNeighborsClassifier(n_neighbors=3)
    ...     scores_k3 = cross_val_score(k3, X, y, cv=10)
    ...     np.mean(scores_k3), np.std(scores_k3)
   Out[30]: (0.96666666666666656, 0.044721359549995787)
   ```

3. Apply the *t*-test to both sets of scores:

   ```
   In [31]: ttest_ind(scores_k1, scores_k3)
   Out[31]: Ttest_indResult(statistic=-0.2873478855663425,
            pvalue=0.77712784875052965)
   ```

As you can see, this is a good example of two classifiers giving different cross-validation scores (96.0% and 96.7%) that turn out to be not significantly different! Because we get a large *p*-value (p=0.777), we expect the two classifiers to be equivalent 77 out of 100 times.

Implementing McNemar's test

A more advanced statistical technique is **McNemar's test**. This test can be used on paired data to determine whether there are any differences between the two samples. As in the case of the *t*-test, we can use McNemar's test to determine whether two models give significantly different classification results.

McNemar's test operates on pairs of data points. This means that we need to know, for both classifiers, how they classified each data point. Based on the number of data points that the first classifier got right but the second got wrong and vice versa, we can determine whether the two classifiers are equivalent.

Let's assume the preceding Model A and Model B were applied to the same five data points. Whereas Model A classified every data point correctly (denoted with a 1), Model B got all of them wrong (denoted with a 0):

```
In [33]: scores_a = np.array([1, 1, 1, 1, 1])
    ...  scores_b = np.array([0, 0, 0, 0, 0])
```

McNemar's test wants to know two things:

- How many data points did Model A get right but Model B get wrong?
- How many data points did Model A get wrong but Model B get right?

We can check which data points Model A got right but Model B got wrong as follows:

```
In [34]: a1_b0 = scores_a * (1 - scores_b)
    ...  a1_b0
Out[34]: array([1, 1, 1, 1, 1])
```

Of course, this applies to all of the data points. The opposite is true for the data points that Model B got right and Model A got wrong:

```
In [35]: a0_b1 = (1 - scores_a) * scores_b
    ...  a0_b1
Out[35]: array([0, 0, 0, 0, 0])
```

Feeding these numbers to McNemar's test should return a small *p*-value because the two classifiers are obviously different:

```
In [36]: mcnemar_midp(a1_b0.sum(), a0_b1.sum())
Out[36]: 0.03125
```

And it does!

We can apply McNemar's test to a more complicated example, but we cannot operate on cross-validation scores anymore. The reason is that we need to know the classification result for every data point, not just an average. Hence, it makes more sense to apply McNemar's test to the leave-one-out cross-validation.

Going back to *k*-NN with *k=1* and *k=3*, we can calculate their scores as follows:

```
In [37]: scores_k1 = cross_val_score(k1, X, y, cv=LeaveOneOut())
    ...  scores_k3 = cross_val_score(k3, X, y, cv=LeaveOneOut())
```

The number of data points that one of the classifiers got right but the other got wrong are as follows:

```
In [38]: np.sum(scores_k1 * (1 - scores_k3))
```

```
Out[38]: 0.0
In [39]: np.sum((1 - scores_k1) * scores_k3)
Out[39]: 0.0
```

We got no differences whatsoever! Now it becomes clear why the *t*-test led us to believe that the two classifiers are identical. As a result, if we feed the two sums into McNemar's test function, we get the largest possible p value, p=1.0:

```
In [40]: mcnemar_midp(np.sum(scores_k1 * (1 - scores_k3)),
    ...:             np.sum((1 - scores_k1) * scores_k3))
Out[40]: 1.0
```

Now that we know how to assess the significance of our results, we can take the next step and improve the model's performance by tuning its hyperparameters.

Tuning hyperparameters with grid search

The most commonly used tool for hyperparameter tuning is **grid search**, which is basically a fancy term for saying we will try all possible parameter combinations with a for loop.

Let's have a look at how that is done in practice.

Implementing a simple grid search

Returning to our *k*-NN classifier, we find that we have only one hyperparameter to tune: *k*. Typically, you would have a much larger number of open parameters to mess with, but the *k*-NN algorithm is simple enough for us to manually implement grid search.

Before we get started, we need to split the dataset as we have done before into training and test sets. Here we choose a 75-25 split:

```
In [1]: from sklearn.datasets import load_iris
   ...: import numpy as np
   ...: iris = load_iris()
   ...: X = iris.data.astype(np.float32)
   ...: y = iris.target
In [2]: X_train, X_test, y_train, y_test = train_test_split(
   ...:     X, y, random_state=37
   ...: )
```

Then the goal is to loop over all possible values of k. As we do this, we want to keep track of the best accuracy we observed as well as the value for k that gave rise to this result:

```
In [3]: best_acc = 0.0
   ...      best_k = 0
```

Grid search then looks like an outer loop around the entire train and test procedure:

```
In [4]: import cv2
   ...      from sklearn.metrics import accuracy_score
   ...      for k in range(1, 20):
   ...          knn = cv2.ml.KNearest_create()
   ...          knn.setDefaultK(k)
   ...          knn.train(X_train, cv2.ml.ROW_SAMPLE, y_train)
   ...          _, y_test_hat = knn.predict(X_test)
```

After calculating the accuracy on the test set (`acc`), we compare it to the best accuracy found so far (`best_acc`). If the new value is better, we update our bookkeeping variables and move on to the next iteration:

```
   ...          acc = accuracy_score(y_test, y_test_hat)
   ...          if acc > best_acc:
   ...              best_acc = acc
   ...              best_k = k
```

When we are done, we can have a look at the best accuracy:

```
In [5]: best_acc, best_k
Out[5]: (0.97368421052631582, 1)
```

Turns out, we can get 97.4% accuracy using k = 1.

With more variables, we would naturally extend this procedure by wrapping our code in a nested for loop. However, as you can imagine, this can quickly become computationally expensive.

Understanding the value of a validation set

Following our best practice of splitting the data into training and test sets, we might be tempted to tell people that we have found a model that performs with 97.4% accuracy on the dataset. However, our result might not necessarily generalize to new data. The argument is the same as earlier on in the book when we warranted the train-test split that we need an independent dataset for evaluation.

However, when we implemented grid search in the last section, we used the test set to evaluate the outcome of the grid search and update the hyperparameter k. This means we can no longer use the test set to evaluate the final data! Any model choices made based on the test set accuracy would leak information from the test set into the model.

One way to resolve this data is to split the data again and introduce what is known as a **validation set**. The validation set is different from the training and test set and is used exclusively for selecting the best parameters of the model. It is a good practice to do all exploratory analysis and model selection on this validation set and keep a separate test set, which is only used for the final evaluation.

In other words, we should end up splitting the data into three different sets:

- a training set, which is used to build the model
- a validation set, which is used to select the parameters of the model
- a test set, which is used to evaluate the performance of the final model

Such a three-way split is illustrated in the following figure:

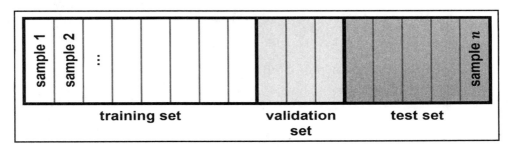

Example of how to split a dataset into training, validation, and test sets

In practice, the three-way split is achieved in two steps:

1. Split into two chunks: one that contains training and validation sets and another that contains the test set:

   ```
   In [6]: X_trainval, X_test, y_trainval, y_test =
      ...:     train_test_split(X, y, random_state=37)
   In [7]: X_trainval.shape
   Out[7]: (112, 4)
   ```

2. Split X_trainval again into proper training and validation sets:

   ```
   In [8]: X_train, X_valid, y_train, y_valid = train_test_split(
      ...:     X_trainval, y_trainval, random_state=37
   ```

```
...      )
In [9]: X_train.shape
Out[9]: (84, 4)
```

Then we repeat the manual grid search from the preceding code, but this time, we will use the validation set to find the best *k* (see code highlights):

```
In [10]: best_acc = 0.0
    ...      best_k = 0
    ...      for k in range(1, 20):
    ...          knn = cv2.ml.KNearest_create()
    ...          knn.setDefaultK(k)
    ...          knn.train(X_train, cv2.ml.ROW_SAMPLE, y_train)
    ...          _, y_valid_hat = knn.predict(X_valid)
    ...          acc = accuracy_score(y_valid, y_valid_hat)
    ...          if acc >= best_acc:
    ...              best_acc = acc
    ...              best_k = k
    ...      best_acc, best_k
Out[10]: (1.0, 7)
```

We now find that a 100% validation score (`best_acc`) can be achieved with *k*=7 (`best_k`)! However, recall that this score might be overly optimistic. To find out how well the model really performs, we need to test it on held-out data from the test set.

In order to arrive at our final model, we can use the value for *k* we found during grid search and re-train the model on both the training and validation data. This way, we used as much data as possible to build the model while still honoring the train-test split principle.

This means we should retrain the model on X_trainval, which contains both the training and validation sets and score it on the test set:

```
In [25]: knn = cv2.ml.KNearest_create()
    ...      knn.setDefaultK(best_k)
    ...      knn.train(X_trainval, cv2.ml.ROW_SAMPLE, y_trainval)
    ...      _, y_test_hat = knn.predict(X_test)
    ...      accuracy_score(y_test, y_test_hat), best_k
Out[25]: (0.94736842105263153, 7)
```

With this procedure, we find a formidable score of 94.7% accuracy on the test set. Because we honored the train-test split principle, we can now be sure that this is the performance we can expect from the classifier when applied to novel data. It is not as high as the 100% accuracy reported during validation, but it is still a very good score!

Combining grid search with cross-validation

One potential danger of the grid search we just implemented is that the outcome might be relatively sensitive to how exactly we split the data. After all, we might have accidentally chosen a split that put most of the easy-to-classify data points in the test set, resulting in an overly optimistic score. Although we would be happy at first, as soon as we tried the model on some new held-out data, we would find that the actual performance of the classifier is much lower than expected.

Instead, we can combine grid search with cross-validation. This way, the data is split multiple times into training and validation sets, and cross-validation is performed at every step of the grid search to evaluate every parameter combination.

The entire process is illustrated in the following figure:

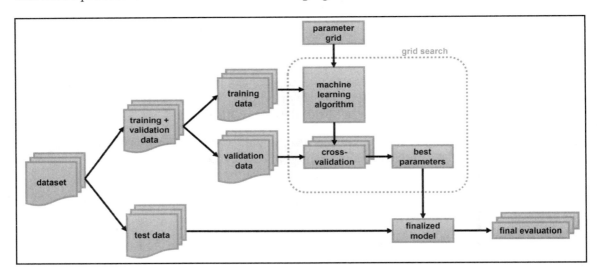

Overview of grid search with cross-validation

As is evident from the figure, the test set is created right from the outset and kept separate from the grid search. The remainder of the data is then split into training and validation sets like we did before and passed to the grid search. Within the grid search box, we perform cross-validation on every possible combination of parameter values to find the best model. The selected parameters are then used to build a final model, which is evaluated on the test set.

Because grid search with cross-validation is such a commonly used method for hyperparameter tuning, scikit-learn provides the `GridSearchCV` class, which implements it in the form of an estimator.

We can specify all the parameters we want `GridSearchCV` to search over by using a dictionary. Every entry of the dictionary should be of the form {name: values}, where name is a string that should be equivalent to the parameter name usually passed to the classifier, and values is a list of values to try.

For example, in order to search for the best value of the parameter `n_neighbors` of the `KNeighborsClassifier` class, we would design the parameter dictionary as follows:

```
In [12]: param_grid = {'n_neighbors': range(1, 20)}
```

Here, we are searching for the best k in the range [1, 19].

We then need to pass the parameter grid as well as the classifier (`KNeighborsClassifier`) to the `GridSearchCV` object:

```
In [13]: from sklearn.neighbors import KNeighborsClassifier
    ...: from sklearn.model_selection import GridSearchCV
    ...: grid_search = GridSearchCV(KNeighborsClassifier(), param_grid,
    ...:                            cv=5)
```

Then we can train the classifier using the `fit` method. In return, scikit-learn will inform us about all the parameters used in the grid search:

```
In [14]: grid_search.fit(X_trainval, y_trainval)
Out[14]: GridSearchCV(cv=5, error_score='raise',
         estimator=KNeighborsClassifier(algorithm='auto',
             leaf_size=30, metric='minkowski',
             metric_params=None, n_jobs=1, n_neighbors=5,
             p=2, weights='uniform'),
         fit_params={}, iid=True, n_jobs=1,
         param_grid={'n_neighbors': range(1, 20)},
         pre_dispatch='2*n_jobs',
         refit=True, return_train_score=True, scoring=None,
         verbose=0)
```

This will allow us to find the best validation score and the corresponding value for *k*:

```
In [15]: grid_search.best_score_, grid_search.best_params_
Out[15]: (0.9642857142857143, {'n_neighbors': 3})
```

We thus get a validation score of 96.4% for *k = 3*. Since grid search with cross-validation is more robust than our earlier procedure, we would expect the validation scores to be more realistic than the 100% accuracy we found before.

However, from the previous section, we know that this score might still be overly optimistic, so we need to score the classifier on the test set instead:

```
In [30]: grid_search.score(X_test, y_test)
Out[30]: 0.97368421052631582
```

And to our surprise, the test score is even better.

Combining grid search with nested cross-validation

Although grid search with cross-validation makes for a much more robust model selection procedure, you might have noticed that we performed the split into training and validation set still only once. As a result, our results might still depend too much on the exact training-validation split of the data.

Instead of splitting the data into training and validation set once, we can go a step further and use multiple splits for cross-validation. This will result in what is known as **nested cross-validation**, and the process is illustrated in the following figure:

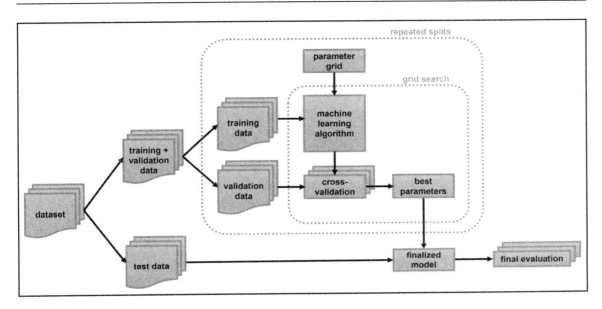

Overview of nested cross-validation

In nested cross-validation, there is an outer loop over the grid search box that repeatedly splits the data into training and validation sets. For each of these splits, a grid search is run, which will report back a set of best parameter values. Then, for each outer split, we get a test score using the best settings.

 While running a grid search over many parameters and on large datasets can be computationally challenging, it is also embarrassingly parallel. This model using a particular parameter setting on a particular cross-validation split can be done completely independently from the other parameter settings and models. This makes grid search and cross-validation ideal candidates for parallelization over multiple CPU cores or over a cluster.

Now that we know how to find the best parameters of a model, let's take a closer look at the different evaluation metrics which we can use to score a model.

Scoring models using different evaluation metrics

So far, we have evaluated classification performance using accuracy (the fraction of correctly classified samples) and regression performance using R^2. However, these are only two of the many possible ways to summarize how well a supervised model performs on a given dataset. In practice, these evaluation metrics might not be appropriate for our application, and it is important to choose the right metric when selecting between models and adjusting parameters.

When selecting a metric, we should always have the end goal of the machine learning application in mind. In practice, we are usually interested not just in making accurate predictions but also in using these predictions as part of a larger decision-making process. For example, minimizing false positives might be equally important as maximizing accuracy.

When choosing a model or adjusting parameters, we should pick the model or parameter values that have the most positive influence on the business metric.

Choosing the right classification metric

We talked about a number of essential scoring functions in Chapter 3, *First Steps in Supervised Learning*. Among the most fundamental metrics for classification were the following:

- **Accuracy**: This counts the number of data points in the test set that have been predicted correctly and returns that number as a fraction of the test set size (sklearn.metrics.accuracy_score). This is the most basic scoring function for classifiers, and we have made extensive use of it throughout this book.
- **Precision**: This describes the ability of a classifier not to label a positive sample as a negative (sklearn.metrics.precision_score).
- **Recall** (or **sensitivity**): This describes the ability of a classifier to retrieve all the positive samples (sklearn.metrics.recall_score).

Although precision and recall are important measures, looking at only one of them will not give us a good idea of the big picture. One way to summarize the two measures is known as the **f-score** or **f-measure** (sklearn.metrics.f1_score), which computes the harmonic mean of precision and recall as *2(precision x recall) / (precision + recall)*.

Sometimes we need to do more than maximize accuracy. For example, if we are using machine learning in a commercial application, then the decision-making should be driven by the business goals. One of these goals might be to guarantee at least 90% recall. The challenge then becomes to develop a model that still has reasonable accuracy while satisfying all secondary requirements. Setting goals like this is often called **setting the operating point**.

However, when developing a new system, it is often not clear what the operating point should be. For this reason and to understand the problem better, it is instructive to investigate all possible trade-offs of precision and recall at once. This is possible using a tool called the **precision-recall curve** (sklearn.metrics.precision_recall_curve).

Another commonly used tool to analyze the behavior of classifiers is the **receiver operating characteristic** (**ROC**) curve. Similar to the precision-recall curve, the ROC curve considers all possible thresholds for a given classifier, but instead of reporting precision and recall, it shows the **false positive rate** against the **true positive rate**.

Choosing the right regression metric

Evaluation for regression can be done in similar detail as we did for classification. In Chapter 3, *First Steps in Supervised Learning*, we also talked about some fundamental metrics for regression:

- **Mean squared error**: The most commonly used error metric for regression problems is to measure the squared error between predicted and true target value for every data point in the training set, averaged across all data points (sklearn.metrics.mean_squared_error).
- **Explained variance**: A more sophisticated metric is to measure to what degree a model can explain the variation or dispersion of the test data (sklearn.metrics.explained_variance_score). Often, the amount of explained variance is measured using the correlation coefficient.
- The R^2 **score** (pronounced **R squared**): This is closely related to the explained variance score, but it uses an unbiased variance estimation (sklearn.metrics.r2_score). It is also known as the **coefficient of determination**.

In most applications we have encountered so far, using the default R^2 score is enough.

As we combine elaborate grid searchers with sophisticated evaluation metrics, our model selection code might become increasingly complex. Fortunately, scikit-learn offers a way to simplify model selection with a helpful construct known as a **pipeline**.

Chaining algorithms together to form a pipeline

Most machine learning problems we have discussed so far consist of at least a preprocessing step and a classification step. The more complicated the problem, the longer this **processing chain** might get. One convenient way to glue multiple processing steps together and even use them in grid search is by using the `Pipeline` class from scikit-learn.

Implementing pipelines in scikit-learn

The `Pipeline` class itself has a `fit`, a `predict`, and a `score` method, which behave just like any other estimator in scikit-learn. The most common used case of the `Pipeline` class is to chain different preprocessing steps together with a supervised model like a classifier.

Let's return to the breast cancer dataset from Chapter 5, *Using Decision Trees to Make a Medical Diagnosis*. Using scikit-learn, we import the dataset and split it into training and test sets:

```
In [1]: from sklearn.datasets import load_breast_cancer
   ...      import numpy as np
   ...      cancer = load_breast_cancer()
   ...      X = cancer.data.astype(np.float32)
   ...      y = cancer.target
In [2]: X_train, X_test, y_train, y_test = train_test_split(
   ...          X, y, random_state=37
   ...      )
```

Instead of the *k*-NN algorithm, we could fit a **support vector machine (SVM)** to the data:

```
In [3]: from sklearn.svm import SVC
   ...      svm = SVC()
   ...      svm.fit(X_train, y_train)
Out[3]: SVC(C=1.0, cache_size=200, class_weight=None, coef0=0.0,
            decision_function_shape=None, degree=3, gamma='auto',
            kernel='rbf', max_iter=-1, probability=False,
```

```
                  random_state=None, shrinking=True, tol=0.001,
                  verbose=False)
```

Without straining our brains too hard, this algorithm achieves an accuracy score of 65%:

```
In [4]: svm.score(X_test, y_test)
Out[4]: 0.65034965034965031
```

Now if we wanted to run the algorithm again using some preprocessing step (for example, by scaling the data first with `MinMaxScaler`), we would do the preprocessing step by hand and then feed the preprocessed data into the classifiers `fit` method.

An alternative is to use a pipeline object. Here, we want to specify a list of processing steps, where each step is a tuple containing a name (any string of our choosing) and an instance of an estimator:

```
In [5]: from sklearn.pipeline import Pipeline
   ...  pipe = Pipeline([("scaler", MinMaxScaler(), ("svm", SVC())])
```

Here, we created two steps: the first, called `"scaler"`, is an instance of `MinMaxScaler`, and the second, called `"svm"`, is an instance of `SVC`. Now we can fit the pipeline like any other scikit-learn estimator:

```
In [6]: pipe.fit(X_train, y_train)
   ...  Pipeline(steps=[('scaler', MinMaxScaler(copy=True,
   ...           feature_range=(0, 1))), ('svm', SVC(C=1.0,
   ...           cache_size=200, class_weight=None, coef0=0.0,
   ...           decision_function_shape=None, degree=3, gamma='auto',
   ...           kernel='rbf', max_iter=-1, probability=False,
   ...           random_state=None, shrinking=True, tol=0.001,
   ...           verbose=False))])
```

Here, the `fit` method first calls `fit` on the first step (the scaler), then it transforms the training data using the scaler, and finally it fits the SVM with the scaled data.

And voila! When we score the classifier on the test data, we see a drastic improvement in performance:

```
In [7]: pipe.score(X_test, y_test)
Out[7]: 0.95104895104895104
```

Calling the `score` method on the pipeline first transforms the test data using the scaler and then calls the `score` method on the SVM using the scaled test data. And scikit-learn did all this with only four lines of code!

Selecting the Right Model with Hyperparameter Tuning

The main benefit of using the pipeline, however, is that we can now use this single estimator in `cross_val_score` or `GridSearchCV`.

Using pipelines in grid searches

Using a pipeline in a grid search works the same way as using any other estimator.

We define a parameter grid to search over and construct a `GridSearchCV` from the pipeline and the parameter grid. When specifying the parameter grid, there is, however, a slight change. We need to specify for each parameter which step of the pipeline it belongs to. Both parameters that we want to adjust, `C` and `gamma`, are parameters of `SVC`, the second step. In the preceding section, we gave this step the name `"svm"`. The syntax to define a parameter grid for a pipeline is to specify for each parameter the step name, followed by __ (a double underscore), followed by the parameter name.

Hence, we would construct the parameter grid as follows:

```
In [8]: param_grid = {'svm__C': [0.001, 0.01, 0.1, 1, 10, 100],
   ...                'svm__gamma': [0.001, 0.01, 0.1, 1, 10, 100]}
```

With this parameter grid, we can use `GridSearchCV` as usual:

```
In [9]: grid = GridSearchCV(pipe, param_grid=param_grid, cv=10)
   ...  grid.fit(X_train, y_train);
```

The best score in the grid is stored in `best_score_`:

```
In [10]: grid.best_score_
Out[10]: 0.97652582159624413
```

Similarly, the best parameters are stored in `best_params_`:

```
In [11]: grid.best_params_
Out[11]: {'svm__C': 1, 'svm__gamma': 1}
```

But recall that the cross-validation score might be overly optimistic. In order to know the true performance of the classifier, we need to score it on the test set:

```
In [12]: grid.score(X_test, y_test)
Out[12]: 0.965034965034965
```

In contrast to the grid search we did before, now for each split in the cross-validation, `MinMaxScaler` is refit with only the training splits, and no information is leaked from the test split into the parameter search.

This makes it easy to build a pipeline to chain together a whole variety of steps! You can mix and match estimators in the pipeline at will, you just need to make sure that every step in the pipeline provides a `transform` method (except for the last step). This allows an estimator in the pipeline to produce a new representation of the data, which, in turn, can be used as input to the next step.

The `Pipeline` class is not restricted to preprocessing and classification but can, in fact, join any number of estimators together. For example, we could build a pipeline containing feature extraction, feature selection, scaling, and classification, for a total of four steps. Similarly, the last step could be regression or clustering instead of classification.

Summary

In this chapter, we tried to complement our existing machine learning skills by discussing best practices in model selection and hyperparameter tuning. You learned how to tweak the hyperparameters of a model using grid search and cross-validation in both OpenCV and scikit-learn. We also talked about a wide variety of evaluation metrics and how to chain algorithms into a pipeline.

Now you are almost ready to start working on some real-world problems on your own. But before we part ways, the next chapter will give you some tips and tricks on how to approach a machine learning problem in the wild.

12
Wrapping Up

Congratulations! You have just made a big step toward becoming a machine learning practitioner. Not only are you familiar with a wide variety of fundamental machine learning algorithms, you also know how to apply them to both supervised and unsupervised learning problems.

Before we part ways, I want to give you some final words of advice, point you toward some additional resources, and give you some suggestions on how you can further improve your machine learning and data science skills.

Approaching a machine learning problem

When you see a new machine learning problem in the wild, you might be tempted to jump ahead and throw your favorite algorithm at the problem—perhaps the one you understood best or had the most fun implementing. But knowing beforehand which algorithm will perform best on your specific problem is not often possible.

Instead, you need to take a step back and look at the big picture. Before you get in too deep, you will want to make sure to define the actual problem you are trying to solve. For example, do you already have a specific goal in mind, or are you just looking to do some exploratory analysis and find something interesting in the data? Often, you will start with a general goal, such as detecting spam email messages, making movie recommendations, or automatically tagging your friends in pictures uploaded to a social media platform. However, as we have seen throughout the book, there are often several ways to solve a problem. For example, we have recognized handwritten digits using logistic regression, k-means clustering, and deep learning. Defining the problem will help you ask the right questions and make the right choices along the way.

Wrapping Up

As a rule of thumb, you can use the following five-step procedure to approach machine learning problems in the wild:

1. **Categorize the problem**: This is a two-step process:
    - **Categorize by input**: Simply speaking, if you have labeled data, it's a supervised learning problem. If you have unlabeled data and want to find structure, it's an unsupervised learning problem. If you want to optimize an objective function by interacting with an environment, it's a reinforcement learning problem.
    - **Categorize by output**: If the output of your model is a number, it's a regression problem. If the output of your model is a class (or category), it's a classification problem. If the output of your model is a set of input groups, it's a clustering problem.
2. **Find the available algorithms**: Now that you have categorized the problem, you can identify the algorithms that are applicable and practical to implement using the tools at our disposal. Microsoft has created a handy algorithm cheat sheet that shows which algorithms can be used for which category of problems. Although the cheat sheet is tailored towards the **Microsoft Azure** software, you might find it generally helpful.

> The Machine Learning algorithm cheat sheet PDF (by Microsoft Azure) can be downloaded from `http://aka.ms/MLCheatSheet`.

3. **Implement all of the applicable algorithms (prototyping)**: For any given problem, there are usually a handful of candidate algorithms that could do the job. So how do you know which one to pick? Often, the answer to this problem is not straightforward, so you have to resort to trial and error. Prototyping is best done in two steps:
    - You should aim for a quick and dirty implementation of several algorithms with minimal feature engineering. At this stage, you should mainly be interested in seeing which algorithm behaves better at a coarse scale. This step is a bit like hiring: you're looking for any reason to shorten your list of candidate algorithms. Once you have reduced the list to a few candidate algorithms, the real prototyping begins.

- Ideally, you would want to set up a machine learning pipeline that compares the performance of each algorithm on the dataset using a set of carefully selected evaluation criteria (see Chapter 11, *Selecting the Right Model with Hyper-Parameter Tuning*). At this stage, you should only be dealing with a handful of algorithms, so you can turn your attention to where the real magic lies: feature engineering.

4. **Feature engineering**: Perhaps even more important than choosing the right algorithm is choosing the right features to represent the data. You can read all about feature engineering in Chapter 4, *Representing Data and Engineering Features*.
5. **Optimize hyperparameters**: Finally, you also want to optimize an algorithm's hyperparameters. Examples might include the number of principal components of PCA, the parameter k in the k-nearest neighbor algorithm, or the number of layers and learning rate in a neural network. You can look at Chapter 11, *Selecting the Right Model with Hyper-Parameter Tuning*, for inspiration.

Building your own estimator

In this book, we visited a whole variety of machine learning tools and algorithms that OpenCV provides straight out of the box. And if, for some reason, OpenCV does not provide exactly what we are looking for, we can always fall back on scikit-learn.

However, when tackling more advanced problems, you might find yourself wanting to perform some very specific data processing that neither OpenCV nor scikit-learn provide, or you might want to make slight adjustments to an existing algorithm. In this case, you may want to create your own estimator.

Writing your own OpenCV-based classifier in C++

Since OpenCV is one of those Python libraries that does not contain a single line of Python code under the hood (I'm kidding, but it's close), you will have to implement your custom estimator in C++. This can be done in four steps:

1. Implement a C++ source file that contains the main source code. You need to include two header files, one that contains all core functionality of OpenCV (opencv.hpp) and another that contains the machine learning module (ml.hpp):

```
#include <opencv2/opencv.hpp>
#include <opencv2/ml/ml.hpp>
```

```
#include <stdio.h>
```

Then an estimator class can be created by inheriting from the `StatModel` class:

```
class MyClass : public cv::ml::StatModel
{
    public:
```

Next, you define `constructor` and `destructor` of the class:

```
MyClass()
{
    print("MyClass constructor\n");
}
 ~MyClass() {}
```

Then you also have to define some methods. These are what you would fill in to make the classifier actually do some work:

```
int getVarCount() const
{
    // returns the number of variables in training samples
    return 0;
}

bool empty() const
{
    return true;
}

bool isTrained() const
{
    // returns true if the model is trained
    return false;
}

bool isClassifier() const
{
    // returns true if the model is a classifier
    return true;
}
```

The main work is done in the `train` method, which comes in two flavors (either accepting `cv::ml::TrainData` or `cv:::InputArray` as input):

```
bool train(const cv::Ptr<cv::ml::TrainData>& trainData,
           int flags=0) const
```

```
{
    // trains the model
    return false;
}

bool train(cv::InputArray samples, int layout,
  cv::InputArray responses)
{
    // trains the model
    return false;
}
```

You also need to provide a `predict` method and a `scoring` function:

```
float predict(cv::InputArray samples,
              cv::OutputArray results=cv::noArray(),
              int flags=0) const
{
    // predicts responses for the provided samples
    return 0.0f;
}

float calcError(const cv::Ptr<cv::ml::TrainData>& data,
                bool test, cv::OutputArray resp)
{
    // calculates the error on the training or test dataset
    return 0.0f;
}
};
```

The last thing to do is to include a `main` function that instantiates the class:

```
int main()
{
    MyClass myclass;
    return 0;
}
```

2. Write a **CMake** file called `CMakeLists.txt`:

```
cmake_minimum_required(VERSION 2.8)
project(MyClass)
find_package(OpenCV REQUIRED)
add_executable(MyClass MyClass.cpp)
target_link_libraries(MyClass ${OpenCV_LIBS})
```

3. Compile the file on the command line by typing the following:

   ```
   $ cmake
   $ make
   ```

4. Run the executable `MyClass` method which was generated by the last command, which should lead to the following output:

   ```
   $ ./MyClass
   MyClass constructor
   ```

Writing your own scikit-learn-based classifier in Python

Alternatively, you can write your own classifier using the scikit-learn library.

You can do this by importing `BaseEstimator` and `ClassifierMixin`. The latter will provide a corresponding `score` method, which works for all classifiers. Optionally, you can overwrite the `score` method to provide your own:

```
In [1]: import numpy as np
   ...: from sklearn.base import BaseEstimator, ClassifierMixin
```

Then you can define a class that inherits from both `BaseEstimator` and `ClassifierMixin`:

```
In [2]: class MyClassifier(BaseEstimator, ClassifierMixin):
   ...:     """An example classifier"""
```

You need to provide a constructor, a `fit` method, and a `predict` method. The constructor defines all parameters that the classifier needs, here given by some arbitrary example parameters `param1` and `param2` that don't do anything:

```
   ...:         def __init__(self, param1=1, param2=2):
   ...:             """Called when initializing the classifier
   ...:
   ...:             Parameters
   ...:             ----------
   ...:             param1 : int, optional, default: 1
   ...:                 The first parameter
   ...:             param2 : int, optional, default: 2
   ...:                 The second parameter
   ...:             """
   ...:             self.param1 = param1
```

```
            self.param2 = param2
```

The classifier should then be fit to data in the `fit` method:

```
        def fit(self, X, y=None):
            """Fits the classifier to data

            Parameters
            ----------
            X : array-like
                The training data, where the first dimension is
                the number of training samples, and the second
                dimension is the number of features.
            y : array-like, optional, default: None
                Vector of class labels

            Returns
            -------
            The fit method returns the classifier object it
            belongs to.
            """
            return self
```

Finally, the classifier should also provide a `predict` method, which will predict the target labels of some data x:

```
        def predict(self, X):
            """Predicts target labels

            Parameters
            ----------
            X : array-like
                Data samples for which to predict the target labels.

            Returns
            -------
            y_pred : array-like
                Target labels for every data sample in `X`
            """
            return np.zeros(X.shape[0])
```

Then you can instantiate the model like any other class:

```
In [3]: myclass = MyClassifier()
```

Wrapping Up

You can then fit the model to some arbitrary data:

```
In [4]: X = np.random.rand(10, 3)
   ...      myclass.fit(X)
Out[4]: MyClassifier(param1=1, param2=2)
```

And then you can proceed to predicting the target responses:

```
In [5]: myclass.predict(X)
Out[5]: array([0., 0., 0., 0., 0., 0., 0., 0., 0., 0.])
```

Implementing a regressor, clustering algorithm, or transformer works similarly, but instead of the ClassifierMixin keyword, you would choose one of the following:

- RegressorMixin if you are writing a regressor (this will provide a basic score method suitable for regressors).
- ClusterMixin if you are writing a clustering algorithm (this will provide a basic fit_predict method suitable for clustering algorithms).
- TransformerMixin if you are writing a transformer (this will provide a basic fit_predict method suitable for transformers). Also, instead of predict, you would implement transform.

This is also a great way to disguise an OpenCV classifier as a scikit-learn estimator. This will allow you to use all of scikit-learn's convenience functions—for example, to make your classifier part of a pipeline—while OpenCV performs the underlying computation.

Where to go from here?

The goal of this book was to introduce you to the world of machine learning and prepare you to become a machine learning practitioner. Now that you know everything about the fundamental algorithms, you might want to investigate some topics in more depth.

Although it is not necessary to understand all the details of all the algorithms we implemented in this book, knowing some of the theory behind them might just make you a better data scientist.

If you are looking for a more advanced lecture, then you might want to consider some of the following classics:

- Stephen Marsland, *Machine Learning: An Algorithmic Perspective. Second Edition*, Chapman and Hall/Crc, ISBN 978-146658328-3, 2014

- Christopher M. Bishop, *Pattern Recognition and Machine Learning*. Springer, ISBN 978-038731073-2, 2007

- Trevor Hastie, Robert Tibshirani, and Jerome Friedman, *The Elements of Statistical Learning: Data Mining, Inference, and Prediction. Second Edition*, Springer, ISBN 978-038784857-0, 2016

When it comes to software libraries, we already learned about two essential ones—OpenCV and scikit-learn. Often, using Python is great for trying out and evaluating models, but larger web services and applications are more commonly written in Java or C++. For example, the C++ package is **vowpal wabbit** (**vw**), which comes with its own command-line interface. For running machine learning algorithms on a cluster, people often use `mllib`, a **Scala** library built on top of **Spark**. If you are not married to Python, you might also consider using R, another common language of data scientists. R is a language designed specifically for statistical analysis and is famous for its visualization capabilities and the availability of many (often highly specialized) statistical modeling packages.

No matter which software you choose going forward, I guess the most important advice is to keep practicing your skills. But you already knew that. There are a number of excellent datasets out there that are just waiting for you to analyze them:

- Throughout this book, we made great use of the example datasets that are built in to scikit-learn. In addition, scikit-learn provides a way to load datasets from external services, such as `mldata.org`. Refer to `http://scikit-learn.org/stable/datasets/index.html` for more information.
- Kaggle is a company that hosts a wide range of datasets as well as competitions on their website, `http://www.kaggle.com`. Competitions are often hosted by a variety of companies, nonprofit organizations, and universities, and the winner can take home some serious monetary prizes. A disadvantage of competitions is that they already provide a particular metric to optimize and usually a fixed, preprocessed dataset.

Wrapping Up

- The OpenML platform (http://www.openml.org) hosts over 20,000 datasets with over 50,000 associated machine learning tasks.
- Another popular choice is the UC Irvine machine learning repository (http://archive.ics.uci.edu/ml/index.php), hosting over 370 popular and well-maintained datasets through a searchable interface.

Finally, if you are looking for more example code in Python, a number of excellent books nowadays come with their own GitHub repository:

- Jake VanderPlas, *Python Data Science Handbook: Essential Tools for Working with Data*. O'Reilly, ISBN 978-149191205-8, 2016, https://github.com/jakevdp/PythonDataScienceHandbook

- Andreas Muller and Sarah Guido, *Introduction to Machine Learning with Python: A Guide for Data Scientists*. O'Reilly, ISBN 978-144936941-5, 2016, https://github.com/amueller/introduction_to_ml_with_python

- Sebastian Raschka, *Python Machine Learning*. Packt, ISBN 978-178355513-0, 2015, https://github.com/rasbt/python-machine-learning-book

Summary

In this book, we covered a lot of theory and practice.

We discussed a wide variety of fundamental machine learning algorithms, be it supervised or unsupervised, illustrated best practices as well as ways to avoid common pitfalls, and we touched upon a variety of commands and packages for data analysis, machine learning, and visualization.

If you made it this far, you have already made a big step toward machine learning mastery. From here on out, I am confident you will do just fine on your own.

All that's left to say is farewell! I hope you enjoyed the ride; I certainly did.

Index

A

accuracy
  about 332
  used, for scoring classifiers 50
activation function 231
AdaBoost
  implementing 295
  implementing, in OpenCV 295
  implementing, in scikit-learn 297
adaptive boosting algorithm 128
additive mixing 104
affinity propagation 211
agglomerative hierarchical clustering
  about 225
  implementing 226
aggregation 86
algorithms
  chaining 334
Anaconda 16
Apache SpamAssassin public corpus
  reference 190
application programming interface (API) 29
artificial intelligence 9
ASCII table
  reference 121
attributes 49
averaging ensembles 272
averaging methods 270
axon 230
axon terminals 230

B

backpropagation
  multi-layer perceptrons, training with 248
bag of words 196
bagging classifier
  implementing 272, 273
bagging methods 272
bagging regressor
  implementing 274
BaggingClassifier class
  options 273
base class 49
basorexia 117
batch-based k-means 215
Bayes' theorem 176, 178, 179
Bayesian classifier
  implementing 181
Bayesian models
  reference, for example 176
bell curve 154, 251
binarizing 88
binary classification 48
binary classification task 144
binary encoding 100
Binder project
  reference 15
blending methods 270
blobs 154
boosting classifier
  implementing 276
boosting ensembles 275
boosting methods 270
boosting regressor
  implementing 277
bootstrapping
  about 166, 316
  implementing, manually in OpenCV 317
  used, for estimating robustness 316
Boston housing prices dataset 70
Breast Cancer Wisconsin dataset
  reference 132
BRIEF descriptor 111

C

C++
  OpenCV based classifier, writing in 341
Caffe
  reference 256
categorical features 100
categorical label 12
categorical variables
  representing 100, 101
class labels
  predicting, classification models used 58
classes 49
classification 48
classification metric
  accuracy 332
  explained variance 333
  precision 332
  R squared 333
  recall 332
  selecting 332
classification models
  used, for predicting class labels 58
classification performance
  measuring 50, 52, 54
classification problem 12
classification task 119
classifiers
  about 270
  scoring, accuracy used 50
  scoring, precision used 50
  scoring, recall used 51
cluster analysis 200
clusters
  organizing, as hierarchical tree 225
coefficient of determination 55
cognitive load 14
Cognitive Toolkit
  about 256
  reference 256
color palette
  reducing, k-means used 220
color pickers 106
color spaces, compressing with k-means
  about 215

true-color palette, visualizing 215, 217
color spaces
  using 104
Compute Unified Device Architecture (CUDA) 14
computer vision 9
conda environment
  OpenCV, installing in 19
Conda-Forge channel
  reference 19
conditional probability 178
confusion matrix 52, 224
continuous outcomes
  predicting, regression models used 68
continuous random variable 174
Continuum
  reference 17
convolutional neural networks 12
core layers, Keras
  Activation 257
  Dense 257
  Reshape 257
corners
  detecting, in images 106
cross-validation procedures
  reference 316
cross-validation
  about 310, 312
  grid search, combining with 328
  manual implementation, in OpenCV 313
curse of dimensionality 92
Cython 15

D

dask 15
data cleaning 85
data formatting 85
data mining 11
data preprocessing
  about 85
  feature extraction 28
  feature selection 28
data representation 84
data sampling 85
data transformation 201
data

classifying, with naive Bayes classifier 186
classifying, with normal Bayes classifier 183, 184, 185
perceptron, fitting to 238
preprocessing 123
visualizing, Matplotlib used 38
decision boundaries 144
decision nodes 115
decision rule 127
decision rules 130, 270
decision rules, hand-coding
  disadvantages 11
decision stump 128, 275, 295
decision tree 10
decision trees, for diagnosing breast cancer
  about 132
  dataset, loading 133
  decision tree, building 134, 136, 137
decision trees
  about 114, 115
  basics 127
  building 117
  combining, into random forest 279
  complexity, controlling 131
  constructing 124
  shortcomings 279, 280, 282
  using, for regression 139
decomposition 86
deep learning
  about 256
  layers 258
deep neural net, training with Keras
  about 264
  convolutional neural network, creating 266
  MNIST dataset, preprocessing 265
  model, fitting 267
dendritic tree 230
dendrogram 225
density-based spatial clustering of applications with noise (DBSCAN) 211
dimensionality reduction
  about 91, 200
  Independent Component Analysis (ICA) 97
  Non-negative Matrix Factorization (NMF), implementing 98
  Principal Component Analysis (PCA) 93
discrete 174
discrete features 100
discriminative models 174, 175
divisive hierarchical clustering 225

E

edge directions 161
elbow method 210
emails, classifying with naive Bayes classifier
  about 189
  data matrix, building with Pandas 193
  data, preprocessing 194
  dataset, loading 190, 191
  full dataset, training 195
  n-grams, used for improving results 196
  normal Bayes classifier, training 195
  tf-idf, used for improving results 197
Enrom-Spam dataset
  reference 190
ensemble methods
  about 270
  averaging methods 270
  boosting methods 270
  stacking methods 270
entropy 131
estimator
  building 341
Euclidean distance 75, 87
evaluation metrics
  used, for scoring models 332
evidence 179
expectation step 204
expectation-maximization solution
  implementing 205, 206
expectation-maximization
  about 204, 205
  limitations 207, 208, 209, 210, 211, 212, 213, 215
expected value 320
explained variance
  about 333
  used, for scoring regressors 55
external datasets
  loading, in Python 36

extremely randomized trees
   implementing 288

F

f-measure 333
f-score 333
factor analysis 200
false negative 52
false positive 52
Fast Approximate Nearest Neighbor (FLANN) 226
FAST keypoint detector 111
feature engineering
   about 28, 84
   feature extraction 85
   feature selection 84
   missing data, handling 89
feature extraction 28, 85
feature selection 28, 84
feature space 28, 85
features
   binarizing 88
   normalizing 87
   scaling, to range 88
   significance 128
   standardizing 86
feedforward neural network 243
folds 310
fraction of variance unexplained 57

G

Gaussian function 154
Gaussian mixture models (GMMs)
   about 205
   reference 205
generative models 174, 175
Gini impurity 130
Git
   about 15
   reference 15
gradient descent 244, 245
gray colormap 42
grid search
   combining, with cross-validation 328
   combining, with nested cross-validation 330
   pipelines, using in 336

H

Haar cascade classifiers 295
hand-coded expert system 10
handwritten digits, classifying with k-means
   about 222
   dataset, loading 222
   k-means, running 222
handwritten digits, classifying
   about 260
   MLP, training with OpenCV 263
   MNIST dataset, loading 260, 261
   MNIST dataset, preprocessing 262
hard voting 298
Harris corner detection 106
Harris operator 28, 85
Hewlett-Packard spam dataset
   reference 189
hidden layer 243
hierarchical clustering
   about 225
   agglomerative hierarchical clustering 225
   divisive hierarchical clustering 225
hierarchical tree
   clusters, organizing as 225
high bias 308
high variance 308
histogram of oriented gradients (HOG) 28, 160
HLS space
   images, encoding in 106
hotspots 154
HSV space
   images, encoding in 106
Hue Saturation Lightness (HSL) 104
Hue Saturation Value (HSV) 104
hyperbolic tangent 251
hyperparameter tuning 170
hyperparameter, tuning with grid search
   about 324
   grid search, combining with cross-validation 328
   simple search grid, implementing 324
   value, of validation set 325, 326
hyperplane 69, 153

I

image features 104
image representation 12
images
   corners, detecting in 106
   encoding, in HLS space 106
   encoding, in HSV space 106
   encoding, in RGB space 104
   representing 104
imputation 89
Independent Component Analysis (ICA)
   implementing 97
information gain 131
internal parameters 26
Internet of Things 13
IPython session
   starting 29
IPython
   reference 30
iris species
   classifying, logistic regression used 76
iterative algorithm 70

J

joblib 15
joint probability distribution 175
Jupyter Notebook 14, 17, 20
Jupyter session
   starting 29

K

k-fold cross-validation
   about 310
   scikit-learn, using for 314
k-means clustering
   about 201
   used, for reducing color palette 219, 220
k-means example
   implementing 201, 202, 204
k-nearest neighbors algorithm 59
k-nearest neighbors implementation, in OpenCV
   about 60
   classifier, training 64
   label, predicting of data point 65, 67

   training data, generating 60, 62
Kaggle
   about 347
   reference 347
Keras
   about 257
   core layers 257
   layers 258
   loss functions 259
   optimizers 258
   reference 259
kernel functions 154
kernel trick 154
kernels
   SVM kernels 155

L

L1 norm 75, 87
L1 regularization 75
L2 norm 75, 87
L2 regularization 75
labels 28, 48
lasso regression 75
layers, deep learning
   Dropout 258
   Embedding 258
   GaussainNoise 258
layers, Keras
   convolutional layers 258
   pooling layers 258
leaf node 127
learning rate 246
leave-one-out cross-validation
   implementing 315
likelihood 179, 186
linear combination 77
linear decision boundary 153
linear regression 68
linear regression, for predicting Boston housing
   prices
   about 70
   dataset, loading 71
   model, testing 72, 73
   model, training 71
Ling-Spam corpus

reference 189
list comprehension 119
logistic function 76
logistic regression
  about 76
  binary classification problem, making 78
  classifier, testing 81
  classifier, training 80
  data, inspecting 79
  data, splitting into test sets 80
  data, splitting into training set 80
  training data, loading 77
  used, for classifying iris species 76
loss function 244
loss functions, Keras
  hinge loss 259
  mean squared error 259

M

machine learning model
  setting up, in OpenCV 49
machine learning problem
  approaching 339, 340
machine learning problems, phases
  test phase 27
  training phase 27
machine learning workflow 26, 27
majority voting 298
Manhattan distance 75, 87
margins 144
Matplotlib
  data, visualizing from external dataset 40
  importing 38
  reference 38
  simple plot, producing 38
  used, for visualizing data 38
maximization step 204
maximum a posteriori (MAP) decision rule 181
maximum-margin classifiers 144
McCulloch-Pitts neuron 230, 231, 232
McNemar's test
  implementing 322, 323
mean absolute error (MAE) 140
mean concave points 136
mean squared error (MSE) 140

mean squared error
  about 70, 333
  used, for scoring regressors 54
members 49
meta-estimators 272
methods 49
mini-batch learning 246
misclassification error 145
MNIST dataset 260
model complexity 308
model performance
  measuring, with scoring functions 50
models
  best model, selecting 307, 308, 309
  combining, into voting classifier 298
  evaluating 304
  evaluating, right way 305
  evaluating, wrong way 304
  scoring, evaluation metrics used 332
multi-class classification 48
multiclass classification task 290
multidimensional arrays
  creating 35
multilayer perceptron (MLP)
  about 243
  training, with backpropagation 248
multilayer perceptron implementation, in OpenCV
  about 249
  data, preprocessing 250
  MLP classifier, creating 250
  MLP classifier, customizing 251, 252
  MLP classifier, testing 255
  MLP classifier, training 253
multivariate Gaussian 93

N

n-gram counts 196
naïve Bayes classifier
  data, classifying with 186
negatives 51
nested cross-validation
  grid search, combining with 330
net input 231
Netflix Prize
  reference 278

neural network, training requisites
    cost function 232
    learning rule 233
    training data 232
Non-negative Matrix Factorization (NMF)
    implementing 98
nonlinear decision boundaries
    dealing with 153
nonlinear support vector machines
    implementing 156, 157
normal Bayes classifier
    data, classifying with 183, 184, 185
Not a Number (NAN) 89
null hypothesis 320
Numerical Python (NumPy) 29
NumPy arrays 32
NumPy
    importing 31
    multidimensional arrays, creating 35
    reference 31
    single array elements, accessing by indexing 34

O

object-oriented programming (OOP) 49
objects 49
Olivetti face dataset 290
on-line gradient descent 246
one-hot encoding 100
open-source deep learning frameworks
    Caffe 256
    Cognitive Toolkit 256
    TensorFlow 256
    Theano 257
    Torch 257
OpenCV based classifier
    writing, in C++ 341
OpenCV's TrainData container, in C++
    used, for dealing with data 43
OpenCV
    AdaBoost, implementing in 295
    cross-validation, implementing manually 313
    installation, verifying 19
    installing, in conda environment 19
    machine learning model, setting up 49
    manual implementation, of bootstrapping 317

ML module 22
Principal Component Analysis (PCA),
    implementing in 93, 95
supervised learning 49
OpenML platform
    reference 348
optimal decision boundaries
    learning 144, 145, 146
optimization problem 70
optimizers, Keras
    adaptive moment estimation 258
    root mean square propagation 258
    stochastic gradient descent 258
origin 69
overfit 308
overfitting 74, 165
overfitting, avoiding
    post-pruning 131
    pre-pruning 131

P

p-value 321
Pandas
    about 193
    reference 37
partial derivative 248
partitions 144
pasting methods 272
pedestrians detection, in wild
    about 158
    dataset, obtaining 158, 159
    histogram of oriented gradients (HOG) 160, 162
    model, bootstrapping 166
    negatives, generating 162, 164
    support vector machine, implementing 165
pedestrians
    detecting, in large image 167
perceptron classifier
    evaluating 239
    toy dataset, generating 237
perceptron learning rule 233
perceptron
    about 233, 234
    applying, to data 241, 242
    fitting, to data 238

implementing 236
pipelines
  about 334
  implementing, in scikit-learn 334
  using, in grid search 336
plane 153
positives 51
post-pruning 131
posterior 179
pre-pruning 131
precision-recall 333
precision
  about 332
  used, for scoring classifiers 50
predictions 68
principal component 96
Principal Component Analysis (PCA)
  implementing, in OpenCV 93, 95
prior 179
probability 77, 174
probability density function 175
probability distribution 174, 175
probability mass function 175
probability theory 174
processing chain 334
pruning 131
Python's NumPy package
  used, for dealing with data 31
Python
  external datasets, loading in 36
  Scikit-Learn based classifier, writing in 344

R

R squared
  about 333
  used, for scoring regressors 55
radial basis functions (RBFs) 154
random forest
  about 131
  decision trees, combining into 279
  implementing 284
  implementing, with scikit-learn 286
random forests, for face recognition
  about 290
  dataset, loading 290

dataset, preprocessing 291
  random forest, testing 293
  random forest, training 293
random number generator 51
random patches methods 272
random subspace methods 272
random variable 174
range
  features, scaling to 88
recall
  about 332
  used, for scoring classifiers 51
receiver operating characteristic (ROC) 333
rectilinear distance 75
recurrent neural network 243
region of interest (ROI) 163
regression metric
  mean squared error 333
  selecting 333
regression models
  used, for predicting continuous outcomes 68
regression problem 12
regression
  about 48
  decision trees, using for 139
regressors
  about 270
  scoring, explained variance used 55
  scoring, mean squared error used 54
  scoring, R squared used 55
regularization 75
regularizors
  L1 regularization 75
  L2 regularization 75
reinforcement learning 13
ResearchGate
  reference 167
results
  significance, assessing 319
reward signal 13
RGB space
  images, encoding in 104
ridge regression 75
robustness
  enabling, bootstrapping used 316

root node 127

S

Scala 347
scale-invariant feature transform (SIFT) 28
Scale-Invariant Feature Transform (SIFT)
  about 84
  using 108
scaling 85
Scientific Python (SciPy) 29
Scikit-Learn based classifier
  writing, in Python 344
scikit-learn
  about 14, 17, 36
  AdaBoost, implementing in 297
  pipelines, implementing in 334
  random forest, implementing with 286
  reference 103, 347
  using, for k-fold cross-validation 314
SciPy 16
scoring functions
  used, for measuring model performance 50
Shi-Tomasi corner detection 107
sigmoid 76
Silhouette analysis 211
silhouette plot 212
simple linear regression 68
snake distance 75
soft voting 298
soma 230
Spark 347
spectral clustering 213
speeded up robust features (SURF) 28
Speeded Up Robust Features (SURF)
  about 85
  using 110
squared error 54
squashing function 77
stacking ensembles 278
stacking methods 270
standardization 86
statistical learning models 49
stochastic gradient descent 246
Student's t-test
  implementing 320, 322

supervised learning
  about 12, 48
  classification 48
  in OpenCV 49
  regression 48
support vector machine implementation
  about 146
  dataset, generating 147
  dataset, preprocessing 148
  dataset, visualizing 147
  decision boundary, visualizing 150, 151, 152
support vector machine
  building 149
SVM kernels 155

T

target labels 119
target values 48
TensorFlow
  about 256
  reference 256
term frequency-inverse document frequency (TF-IDF)
  about 103
  reference 103
term–inverse document frequency (tf–idf) 197
test data 12
test dataset 27
text features
  representing 102
Theano
  about 257
  reference 257
thresholding 88
Torch
  about 257
  reference 257
toy dataset
  creating 181
trained decision tree
  visualizing 125
training data 12
training dataset 27
true negative 52
true positive 52

true-color image 215
two-class classification 48
type I error 52
type II error 52

U

UC Irvine machine learning repository
  reference 348
uncertainty 131
underfit 308
universal approximation property 244
unsupervised learning
  about 12, 200
  cluster analysis 200
  dimensionality reduction 200
  factor analysis 200

V

validation dataset 27
validation set 326
variance 57
variance reduction 140
version control system 15
voting classifier
  about 298
  implementing 299
  models, combining into 298
voting schemes
  hard voting 298
  soft voting 298
vowpal wabbit (vw) 347

W

weight coefficients 68, 231
weight update rule 233
weighted sum 231
wisdom of the crowd concept 270